The Science of Subjectivity

Joseph Neisser
Grinnell College, USA

© Joseph Neisser 2015

All rights reserved. No reproduction, copy or transmission of this publication may be made without written permission.

No portion of this publication may be reproduced, copied or transmitted save with written permission or in accordance with the provisions of the Copyright, Designs and Patents Act 1988, or under the terms of any licence permitting limited copying issued by the Copyright Licensing Agency, Saffron House, 6–10 Kirby Street, London EC1N 8TS.

Any person who does any unauthorized act in relation to this publication may be liable to criminal prosecution and civil claims for damages.

The author has asserted his right to be identified as the author of this work in accordance with the Copyright, Designs and Patents Act 1988.

First published 2015 by
PALGRAVE MACMILLAN

Palgrave Macmillan in the UK is an imprint of Macmillan Publishers Limited, registered in England, company number 785998, of Houndmills, Basingstoke, Hampshire RG21 6XS.

Palgrave Macmillan in the US is a division of St Martin's Press LLC, 175 Fifth Avenue, New York, NY 10010.

Palgrave Macmillan is the global academic imprint of the above companies and has companies and representatives throughout the world.

Palgrave® and Macmillan® are registered trademarks in the United States, the United Kingdom, Europe and other countries.

ISBN: 978–1–137–46661–7

This book is printed on paper suitable for recycling and made from fully managed and sustained forest sources. Logging, pulping and manufacturing processes are expected to conform to the environmental regulations of the country of origin.

A catalogue record for this book is available from the British Library

A catalog record for this book is available from the Library of Congress.

For my parents, Arden and Ulric

Contents

Acknowledgments		viii
1	Introduction: Consciousness, Subjectivity, and the History of the Organism	1

Part I Subjectivity Considered as the First-Person Perspective

2	Subjectivity and Reference	11
3	Unconscious Subjectivity	38
4	What Subjectivity Is Not	61

Part II Subjectivity in the Neurobiological Image

5	Interlude: The Neurobiological Image	83
6	The Science of Subjectivity: Neurobiology and Evolutionary Development	87
7	Putting the *Neuro* in Neurophenomenology	110
8	Neural Correlates of Consciousness Reconsidered	140
Postscript: Neurophilosophy, Darwinian Naturalism, and Subjectivity		157
Notes		166
Works Cited		187
Index		201

Acknowledgments

This book was completed with the help of the Harris Fellowship at Grinnell College and the generous offices at the Centre for Subjectivity Research (CFS) at the University of Copenhagen.

I received thoughtful feedback from John Fennell, Johanna Meehan, Dan Zahavi, John Michael, Bernardo Ainbinder, Joona Taipale, Kent McClelland, Jakob Howhy, and several anonymous referees. Any errors are mine.

Thanks most of all to my wife Cora Jakubiak, whose love and friendship made this book possible.

The author and publisher are grateful to the publishers who have granted permission for the following papers or versions thereof to be reprinted here:

Chapter 4 appeared as: What subjectivity is not. *Topoi: An International Review of Philosophy* forthcoming (2015). Available online 14 August 2014. DOI: 10.1007/s11245-014-9256-5.

Chapter 8 appeared as: Neural correlates of consciousness reconsidered. *Consciousness and Cognition* 21 (2012) 681–690.

Parts of Chapter 2 appeared as: Subjectivity and the limits of narrative. *Journal of Consciousness Studies* 15, 2 (2008) 51–66.

An earlier version of Chapter 3 appeared as: Unconscious subjectivity. Psyche: *An Interdisciplinary Journal of Research on Consciousness* 12, 3 (September 2006).

1
Introduction: Consciousness, Subjectivity, and the History of the Organism

This book is inspired by the deeply historical science of the organism found in contemporary biology. Since P.S. Churchland's *Neurophilosophy* (1986), "neuro" has become a familiar prefix. Neuroethics, neuroeconomics, neuroaesthetics, neurophenomenology and the neurodiversity movement are all contemporary expressions of an intellectual zeitgeist. A *neurobiological image* of mind and person is emerging, a picture in which we are hide-bound animal subjects, neurologically enabled, ecologically situated, and historically conditioned. But the true philosophical payoff of the newfangled neuroscience remains unclear. What does it contribute, exactly, to our understanding of human experience? In particular, can it really help us understand *how anything like a first-person perspective could arise in the world?* (Levine, 2001). My answer is yes. The new biology provides real insight into the nature of "for-me" subjectivity and how it is elaborated in the life of an animal. Evolutionary developmental biology offers the framework for an historical account of the way first-person experience arises and how it functions to enable organisms like us to navigate their world. In short, *subjectivity is historically conditioned embodiment.* In what follows I set forth a philosophical analysis of the first-person perspective, its central place in mental life, and its probable neurobiological basis. My aim is to articulate a conceptually powerful and empirically informed account of subjectivity that will be relevant to audiences across the humanities and life sciences.

The content of subjective experience is vast and diverse. Fleeting qualities succeed one another with an "inconceivable rapidity" (Hume). It is notoriously difficult even to describe *what-it-is-like*, let alone to explain it. Nevertheless, experience has a certain minimal structure, consisting in the way things are apprehended first-hand, through a first-person perspective. No matter what life is like, it is always *for-me*. I begin by arguing that the subjectivity of experience – it's first-person aspect – is a matter of its form, not its qualitative content. First-person mental states consist in a structure in which "I" am *specified* by experience but not represented in experience. Roughly, a given representation is for-me insofar as the question *who* does not and cannot arise. This basic idea guides a naturalistic approach to embodied subjectivity firmly grounded in philosophy of biology and psychology. Part One of the manuscript, "Subjectivity considered as the first-person perspective," focuses on articulating an analysis of subjectivity culled from philosophy of mind and cognitive science. Part Two, "Subjectivity in the neurobiological image," moves to identify an empirical framework for explaining subjectivity from the perspective of biology, neurophilosophy, and neurophenomenology.

1 The idea and the approach

The claim that experience essentially involves a subjective point of view has been a constant theme in the literature on consciousness (e.g., Kriegel, 2009; Zahavi, 2005; Levine, 2001; Nagel, 1974). What is needed is an analysis of the relation between subjectivity and consciousness, showing how experiences are "lived" but not necessarily known, doing justice to the idea that the first-person perspective involves a distinctively "direct" kind of representation. I argue that although the first-person perspective is essential for consciousness, the two are not equivalent because subjective thought can take place without awareness. An adequate conception of subjectivity should be capable of extending to a class of for-me mental states that remain unconscious or pre-conscious. In short, while conscious awareness is sufficient for subjectivity, it is not necessary.

It is usual to distinguish representations that are *for-me* from representations that are merely *in me* as information processes. The for-me/in me distinction is supposed to cleanly separate the conscious and subjective arena from the unconscious and "subpersonal" domain of cognition (e.g., Kriegel, 2009). But I argue that the for-me/in me distinction cross-cuts the conscious/unconscious distinction because there is something it is like, e.g., to have an unconscious emotion. If there is something it is like, this must be for-me or first-person. This reveals an inherent tension in standard ways of thinking about subjective consciousness. It is understood as the *presence* of phenomenal quality (what it is like for me), but at the same time it is said to consist in an *awareness* of phenomenal quality (knowledge of what it is like for me). I argue that the former does not require the latter. For-me mental states can exist in me, without my knowledge. Thus, subjectivity runs deeper than conscious awareness (cf. Block, 1995, 2007, 2011). Careful reconsideration of the debate surrounding the so-called *Mesh Argument* in cognitive science indicates that there is an empirical basis for the idea that there are subjective mental states of which the subject is unaware. And while there is good evidence for the existence of unconscious emotions and other affective processes, neuropsychological research is hampered by the assumption that only conscious thought can have a first-person form (e.g., LeDoux, 2004).

In what, then, does subjectivity consist? The point of entry is the notion of *identification-free self-reference* (Evans, 1982). In the first-person or subjective mode of presentation, the question *to whom* the experience is presented cannot arise. Pain is a primary example (Shoemaker, 1981). There is no identifying judgment or inference involved in moving from the experience of pain to "I am in pain." With respect to subjectivity, consciousness of pain is equivalent to the thought "I am in pain." Note that the analysis works for other qualia, too: with respect to the subjective component, the experience of reddishness is equivalent to the thought "I see red." Using identification-free reference as my guide, I argue that the distinctive architecture of the first person consists in the way the "I" of experience is always *absent*. I develop this idea further by showing how for-me representation consists in a holistic or gestalt scheme in which what-it-is-like is always experienced "from the inside." Particular contents are nested within the ongoing global subjective scheme. Thus, subjective experience is apprehended as for-me, such that an "I" is specified by the structure of the experience itself.

Considering subjectivity as the identification-free form of the first-person perspective makes a functional analysis possible. Ecologically considered, subjectivity serves an *orienting* function for an animal in the environment of its concern. Orientation (or coordination) is essential to the notion of identification-free self-reference. The "I" of experience is specified in an identification-free manner in virtue of being specified *only* in relation to the world (Ismael, 2009; Cassam, 1997). Subjective orientation, then, is the animal's valenced sense of directionality in its local environment. But baseline subjective orientation does not require a self that is specifically conceptualized *as* an "I". Nor does it require full-blown reflective consciousness or metacognitive awareness of itself specifically as qualitative or internal content. It need only enact a first-person point of view. First-person subjectivity specifies "where you are coming from" in the practical situation. It is the way an individual is engaged in life, the basis on which things matter. Subjective engagement with the world is sometimes called *coping*.

The search for an explanation of concerned coping leads to evolutionary developmental biology (evo-devo, or the "expanded synthesis"). The developmental function of subjectivity is learning to cope with the environment of concern. Therefore, the basic problem for the science of subjectivity is animal *navigation,* and the model explanation is the *cognitive map*. The first-person perspective partly consists in "finding your bearings;" i.e., in navigation dynamically linked with affective systems. This baseline affective engagement with the world is part of the deep structure of for-me mental states, governing the first-person point of view. Baseline subjectivity in this sense is achieved by species across vertebrate phyla and beyond, irrespective of reflective ability. The baseline structure of the first-person perspective appears as a range of diverse but related neuropsychological character states. Evolutionary developmental biology suggests that this shared structure is rooted in a set of homologous mechanisms, operating under historical and developmental constraints (Jacobs, 2012; Griffiths, 2007). The mechanisms of concerned engagement are also currently investigated within the research program known as "affective neuroscience" (Panksepp, 2012, 2007, 2005, 1998) which, I argue, should be reinterpreted in light of the evo-devo framework. Thus, my approach in *The Science of Subjectivity* is to tackle the problem of subjectivity head on by (1) providing structural and dynamic analyses in terms of the first-person perspective, (2) identifying neuropsychological functions for the structural and dynamic forms, and (3) connecting these functions with underlying mechanisms via the fundamentally historical and comparative methods of evolutionary developmental biology.

The account must do justice to the way subjectivity is contingent on our particular form of embodiment, and this means moving beyond straight functionalism in philosophy of mind. An essential feature of functionalism is the notion of multiple realization: the idea that functions can be carried out in many different ways. For example, a robot vacuum cleaner could navigate the apartment without any attendant subjectivity in performing this function. This illustrates the subtle sense in which (pure) functionalism remains a "disembodied" theory of mind, inadequate to the first-person perspective. Although it holds that every mental state is realized in *some* kind of embodiment, functionalism gives no explanatory weight to these mere "implementation" details. Translating functionalism to a biological context has yielded evolutionary psychology in its mainstream form (Cosmides & Tooby, 1991). According to that approach, each neuropsychological feature is an adaptation, and explaining it consists in (a) identifying a functional role in the contemporary organism, and (b) reconstructing a scenario in which the capacity to perform the function might have been selected for. The evo-devo framework moves beyond evolutionary psychology by incorporating *homology thinking* about structure and lineage (Ereshefsky, 2012). "Homology thinking" names a set of historical and comparative explanatory strategies, used throughout biology, capable of distinguishing the different paths by which biological structures develop, and how these historical differences affect the life of the organism in distinctive ways. I draw on homology thinking to formulate a new approach to embodiment that incorporates elements of both functionalism and embodied holism. In a slogan, embodiment is history.

This explanatory framework bears implications for the search for the Neural Correlates of Consciousness (NCC). I approach key NCC research (in particular, the binocular rivalry paradigm) from the perspective of philosophy of science rather than in the usual way, through the lens of philosophy of mind. Doing so, I argue, reveals that the binocular rivalry experiments are part of a larger mechanistic strategy aimed at manipulating component causes within complex biological systems (Neisser, 2012a; 2012b). Philosophical interpreters of NCC research seriously misrepresent the actual empirical practice by framing it as a search for local neural activity that bears a semantic or logical relation to conscious experience. I argue that NCC research is best characterized as an attempt to locate a *causally relevant* neural mechanism and not as an effort to isolate a discrete neural representation, the content of which correlates with some actual experience. Reflection reveals that the standard conception of the NCC is inadequate precisely because it omits to consider the first-person phenomenology of subjective experience (Noe & Thompson, 2004). The account of subjectivity offered here is more phenomenologically valid because it indicates how isolated sensory representations might become nested within an ongoing, holistically structured first-person context. Thus, the causal interpretation of NCC research that I offer fits well with the evo-devo framework. In this picture, mechanisms identified first through homology and comparative biology are further investigated through manipulation and intervention.

To be "phenomenlogically valid," the account should show how key features of the first-person perspective arise. One such feature, emphasized and carefully analyzed in the Husserlian tradition, is *temporality*.

Subjectivity doesn't just take place *at* a time; it is itself a temporal articulation of intentional content. Husserl identified this temporal dimension of subjectivity and characterized it as a continuous structure of protention and retention, a "direct" intuitive grasp of the about-to-happen and the just-past (1962/1977). The sense of temporal flow does not take its own, separate intentional object (it is not the consciousness of time), but instead is part of the "microstructure" of the first-person perspective (Gallagher & Zahavi, 2012). I argue that temporality is what happens when animals like us navigate the world of concern. The first-person perspective points back, as it were, to a temporal position relative to events. The identification-free structure of subjectivity partly consists in the "here-and-now of things," a subject-position which is specified without being explicitly represented. Thus, subjective temporality *does* something. It orders particular contents within a larger gestalt. By treating temporality as an aspect of animal navigation, I provide an example of how to put the *neuro* in *neurophenomenology*, and I contrast this approach with the enactive account offered by Thompson (2007).

In pursuing this line of inquiry, I also argue against the standard philosophical view about the relation between subjectivity and consciousness. In the recent literature, subjectivity is primarily understood as *direct awareness of qualia*, a conception that renders subjectivity a special problem for philosophy of mind, over and above the problem about qualia. On this basis, Levine (2001) has suggested that the hardest part of explaining consciousness is explaining how anything like a subjective point of view could arise in the natural world. I agree with the "friends of qualia" that we will not have a complete scientific explanation of consciousness as long as the problem of qualia persists. But I argue that inner awareness is not the essence of subjective thought and that the first-person perspective can be treated independently of the problem of qualia. So I reject the core assumption, shared by researchers of all stripes and on both sides of the Atlantic, that subjectivity is inherently conscious. But subjectivity *is* partly constitutive of consciousness, and explaining it therefore contributes to an explanation of consciousness. The neurobiological image presents a portrait of a minimally cognitive first-person perspective that demands attribution of subjectivity to a very wide range of animals. When neurobiological subjectivity is understood as the way animals learn to cope with life, its central importance can also be seen. *The Science of Subjectivity*, then, also provides a naturalistic and phenomenologically valid interpretation of the Knowledge Argument (Jackson, 1982).

2 Relation between this study and others in its field

Within contemporary analytic philosophy of mind, the agenda was set by Levine (2001), who argued that no representational approach, and in particular no "higher-order thought" (HOT) theory of the general kind offered by Rosenthal (2000) or Lycan (2004), can explain subjectivity. This theme has been further pursued in an important monograph by Kriegel, entitled *Subjective Consciousness: A self-representational theory* (2009). Like Levine, Kriegel holds that subjectivity *just is* qualitative awareness. On his view, a mental state is subjectively conscious if and only if it represents itself in the right way. Kriegel's work contrasts with mine in two main ways: (1) He holds that subjectivity is reflexively structured such that the representation comes to take itself as part of its own content, whereas I argue that the distinctive mark of subjectivity is the peculiar *lack* of reflexivity displayed in identification-free self-reference. (2) He equates subjectivity with awareness of what-it-s-like, denying the possibility of unconscious subjectivity. I argue that awareness is not the essence of for-me-ness and that unconscious subjectivity is a fact of life. Thus, my approach is complementary to the HOT theory, which I take to be an account of our *knowledge* of our own subjectivity. Conscious states are subjective states we are aware of having, and I hold that this awareness can be understood as metacognition (Neisser, 2004; Koriat, 2000). A mental state is metacognitive if it is directed at other mental states of the cognizer. To be consciously aware, then, is to know (represent) your own subjectivity.

The Science of Subjectivity also connects with the literature on enactivism and neurophenomenology. Thompson's *Mind in Life* (2007) presents a sophisticated mixture of phenomenology and philosophy of biology, the most complete expression of neurophenomenology to date. Thompson not only questions the received conception of subjectivity, but attempts to rethink it. He formulates the issue in terms of the "body-body problem," the problem of relating one's subjectively lived body to the living body that one is. To address this version of the problem, Thompson articulates the theory of *sensorimotor subjectivity* according to which bodily self-consciousness arises through the dynamic interaction of brain, body, and world. Thompson finds a quite general analogy between mind and life, and uses this to develop a phenomenological version of autopoeisis (self-organization). In contrast, I argue for a specific causal-historical connection between animal navigation and subjective coping. While his paradigm is the biological cell or amoeba, mine is the cognitive map. This also points to a more fundamental difference with Thompson. In the end, his interest in autopoeisis is ontological not explanatory. In particular, Thompson holds that transcendental consciousness precedes "objectivity" and all scientific cognition. The result is a deeply non-naturalistic brand of emergentism. In contrast, I hold that the evo-devo framework promises a true science of the organism that can play a straightforwardly explanatory role in a theory of subjectivity.

Part I
Subjectivity Considered as the First-Person Perspective

2
Subjectivity and Reference

The problem of subjectivity is an explanatory problem, and the problem is not merely that we do not yet have the explanation. We don't seem to have even the framework for an explanation. In Thomas Metzinger's phrase, subjectivity has not been "turned into an explanandum for the hard sciences" (2004, p. 36). We don't know what an explanation would look like, or what sort of explanation might be given. This is because subjectivity seems *sui generis*. There are no explanations of other, relevantly similar phenomena that might serve as paradigms. In part, the difficulty lies in knowing what needs to be explained. Subjectivity is a bit like an "unknown unknown." When confronted with a *known* unknown, there is some known variable x, such that discovering the value of x will fit some model or solve some equation. But here the variable itself is undefined, an unknown. Although it is indeed known *that* we don't have the explanation (as in the case of a "known unknown"), this doesn't help because scientists have little more than the word itself to work with. Like Meno, we don't know what we are looking for. In seeking an explanation of subjectivity, what do we want to know? What is subjectivity? How will we know if we've explained it? The task of the first three chapters of this book is to consider these questions and to clarify what needs an explanation and what does not. Treating subjectivity as the first-person perspective, I sketch an approach according to which the "directness" of experience consists in a distinctive referential structure. Roughly, what makes an experience an experience *of x's* consists in the way it architecturally specifies x. In this chapter I introduce this idea, defend my interpretation of it, and put it to use.

1 The mark of subjectivity: identification-free self-reference

Subjective consciousness is always "first-person" in that it is *for* some subject (Kriegel 2009; Neisser, 2008; Zahavi, 2005; Levin 2001; Nagel, 1974).[1] What does this mean? What is the difference between a structure that is "in the first-person" and one that is not? Some criterion of subjectivity is needed, a mark by which to discriminate experiential structure from other more familiar breeds of representation. The relevant mark of subjectivity is most clearly seen in a peculiar use of the first-person pronoun. Wittgenstein (1958) distinguished between two uses of "I", the *object-use* and the *subject-use*, and this will be the point of entry into the account of the first-person perspective. The unique referential structure of the subject-use of "I" sheds light on the directness of subjective experience, on what it is for experience to be for-me. Please note at the outset that in proceeding this way, I do not claim that subjectivity consists in or requires linguistic mastery of the first-person pronoun. Nor do I mean that every subjective thought is a token of a proposition with the relevant pronoun or concept as a constituent. Rather, first-person experience has an identification-free structure, and the logic of the first-person pronoun is the place in language where this subjectivity can be seen. Subjectivity is a function of the first-person perspective, and the subject-use of "I" is an *illustration* of the structure that characterizes first-person experience. Further, the distinction between the two uses of "I" provides a criterion for theories of subjectivity. If a putative account concerns the other, object-use of the first-person pronoun, then that theory is not an account of subjectivity at all but rather an account of "self" or "person."

In the object-use of "I" the pronoun appears as part of a complex judgment or thought. It involves the conjunction of two independently evaluable claims. In this object-use, the judgment "I am P" combines (a) "x is P" with (b) "I am x." Claim (a) is the *predication* component of the judgment and (b) is the recognition or *identification* component.[2] These two parts of the complex are independent in the sense that the former, predication judgment can be perfectly correct at the same time that the latter, identification component is incorrect. That is, it is possible that "x is P", but I am not x. So, in the object-use of the first-person pronoun, claims that purport to be about "me" can sometimes fail to be because I mistakenly identify myself with some other person. "I am on the giant video screen at the ballpark" can be a false judgment, and it can be false relative to the pronoun it employs. It can perfectly well be the case that *someone* is on the giant video screen, but I am not that person.

By contrast, the subject-use of the first-person pronoun does not involve a complex judgment in this way. It contains no identification component. Rather, this use of the first-person pronoun reflects the from-the-inside architecture of subjective experience. Evans (1982) calls it *identification-free self-reference*. We need not ask *to whom* the predicate applies, because this is not an external fact independent of the predicate judgment itself. Years earlier, Shoemaker put the key point this way:

> There is an important and central class of psychological predicates, let us call them P* predicates, each of which can be known to be instantiated in such a way that knowing it to be instantiated in that way is equivalent to knowing it to be instantiated in oneself. (Shoemaker, 1968 p. 16)

Pain is the primary example of a P* predicate. With respect to the identification component, subjective experience of pain is equivalent to the thought "I am in pain." And this is generally true for all judgments of qualia. For example, with respect to the identification component, the experience of reddishness is equivalent to the thought "I see reddishness." This connection between the structure of the experience and the subject-use of "I" provides a way to think about the *directness* (phenomenological immediacy) of subjective experience which avoids the problems associated with the widespread idea (discussed in Chapter 4) that subjectivity consists in an inner awareness of qualia.

The subject-use of the first-person pronoun (the class of P* predicates) corresponds to the psychological states apprehended from-the-inside. This nicely picks out the phenomenon of primary interest here, namely, the sense in which experience is for-me. It also clarifies the specific, tightly circumscribed sense in which subjectivity has a special epistemic status. Identification-free self-reference displays "immunity to error through misidentification" (the phrase was coined by Shoemaker). Wittgenstein expressed this by saying that, in the subject-use of the first-person pronoun there is no possibility of error, while in the object-use the possibility of an error "has been provided for." But notice just how extremely limited this special epistemic status really is. First, the first-person point of view does not deliver knowledge of the subject's identity, because it contains no identifying content. Instead, the distinctive architecture of the first-person consists in the *absence* of an identification judgment. The immunity involved in the subject-use of "I" consists in the fact that no claim is made.[3] So being immune to error through misidentification does not mean we can get knowledge of "who" we

14 *The Science of Subjectivity*

are by this route. Just as importantly, identification-free self-reference most emphatically does not guarantee that the predicate component of the judgment is correct! This allows for the possibility that one can be mistaken whether one's experience is reddish, even while being immune to error about whether "I" am the subject of the experience. Next I further explore this basic idea, clarify my interpretation of it, and defend it from objection.

1.1 Varieties of reference: descriptive vs. identification free

We need to know a bit more about this peculiar form of reference and how it works. There are several closely related concepts that need to be sorted out. To begin with, identification-free self-reference is not the only kind of reference that can display immunity to error through misidentification. Certain standard kinds of "objective" judgments can also display this property. So what is it about the subject-use of "I" that is different, or distinctively direct? Once this is sorted out, two aspects of more properly identification-free reference can be distinguished: for-me-ness and simple perceptual demonstration.

First, some cases of immunity to error through misidentification are not cases of identification-free self-reference. Russell's theory of descriptions can also handle some expressions with this kind of immunity.[4] On Russell's model, the reference of a proper name or singular term is fixed by a description that uniquely picks out the individual in question. For example, the meaning of the name "Doc" might be fixed by "composer of *Columbus Stockade Blues*." On the basis of this definition, together with other evidence that there was only one composer of the song, you might judge that "Doc is the sole author of *Columbus Stockade Blues*." Clearly this judgment might be mistaken. Contrary to your other evidence, the song might be a traditional folk song. If so, then it might not have a sole composer. But if you are right that *someone* exists who fits the description (i.e., who was the sole composer), then that person can be no other than Doc. This follows from the definition of "Doc." Or again, if *anyone* is the sole composer of *Columbus Stockade Blues*, then – necessarily – Doc is that person. Hence, the judgment is immune to error through misidentification. It is not identification free, but the identification is guaranteed to be *error* free.

The above example also illustrates a general feature of all judgments immune to error through misidentification. In all such cases, if the singular judgment is false then the corresponding existential generalization is also false. In the example, if it is false that Doc is the sole composer of *Columbus Stockade Blues* (Cd) then it is also false that *someone* was the sole composer of *Columbus Stockade Blues*, or $(\exists x)Cx$. By contrast, judgments that are vulnerable to error through misidentification are those in which such a mistake does not entail a mistake about the corresponding existential generalization.[5] Again recall the ballpark video in the first example introduced above: In that case I am vulnerable to error through misidentification since I can still be right that there is someone on the giant video screen, even if I am mistaken that I am that person. But although the judgment that Doc is the sole composer of *Columbus Stockade Blues* is immune to misidentification error, nevertheless it is no more "first-person," "subjective," or "direct," than any other singular proposition. It does not refer to Doc in any distinctive way. So something more than immunity is needed to characterize the for-me-ness of subjectivity, namely, the *way* the immunity is accomplished. The peculiar immunity displayed in the subject-use of "I" arises from its identification-free form, not from any descriptive concept (definition). I take a moment to explore this contrast.

By fixing the reference of a name through a defining description, you can guarantee that the name will track the predicates mentioned in the description. If you then confine your judgment to an application of the identifying description to the name it defines, the judgment will be immune to error through misidentification because it is tailored to assure that its identification component cannot be false. But surely first-person experience is not tailored in this way. It does not consist in "definitional" truths or semantic claims. I cannot be wrong that it is I who has this headache, but this immunity is not grounded in a prior description of me, picking out this headache. To embrace this would be to set oneself on the path of Leibniz. It would be folly to explain subjectivity by appeal to the semantics of a complete concept or total objective description of "my" mental life that includes each and every P* predicate as a necessary part of my essence. Whether or not I have a headache now, or see a grey lake now, or whatever, is an entirely contingent matter of fact. The great thing about first-hand experience is that you can learn so much from it. As Yogi Berra said, you can see a lot just by observing.[6] These non-definitional first-person judgments are immune to error through misidentification but this cannot be accounted for on the model of descriptive names.[7] Some other kind of reference-fixing must underlie the first-person.

What this shows is that the peculiar immunity illustrated by the subject-use of the first-person pronoun is not, strictly, a *semantic* feature.[8] The first-person perspective is not governed by the logic of any concept to which the linguistic form refers, such as "I" or "me." Rather, identification-free self-reference provides a clue to the way we are "hooked-up" or embodied. The right account of how we are embodied in the world will also explain why subjective experience is structured the way it is.[9] Identification-free reference is a matter of the *dedicated* manner in which information is gained, and this is the first-person "mode of presentation." Thus:

> In the case of the first-person...the subject is using ways of finding out about the world that are, as we might say, "dedicated" to the properties of one particular object, namely, that very person. They are not ways of finding out that could be equally well applied to any range of objects. It is for that reason that although the subject using such a way of finding out can make a mistake, it could not be a mistake about who is in question. (Campbell, 1999, p95)

Perceptual information is not for just anyone. Perceptual systems function to *locate* a perceiver in relation to a perceived world. The "properties" of the subject that Campbell mentions in the quote are contingent relational properties specifying the world from here. Most fundamentally, this relation is not tracked by means of a description of the subject and the world, but through the informative *coordination* of the two. Subjective thought involves a kind of pointing, i.e., a kind of demonstrative reference.[10] The world of experience is always the world for-me, but this fact need not be further represented (or self-represented, or meta-represented) for the perceiver. Instead, "I" am directly *specified* in the first-person without being explicitly identified (either tagged or described) therein. The upshot is that subjectivity is not based in a "self" at all. Rather, the first-person perspective is oriented around here.[11]

Roughly then, what makes an experience an experience of x's consists in the way it architecturally specifies x.[12] So far, "architectural specification" simply means that the first-person perspective is identification free. But one further note can be added right away. When one points to an object *there*, one uses its perceived location to fix the reference from here: "The coffee cup is *there*." Now, this thought can certainly be mistaken in various ways. Whatever is there might not be "a coffee cup" but say, a work of art. Or, unbeknownst to me, there could be a series of projectors creating the illusion that the piece of equipment is at that location. Whatever. Those errors are not errors of misidentification. If the object *there* is not a coffee cup, it is still the object there, and not another object. Similarly, if I am laboring under an illusion about the apparent location or existence of the object, it is the location of *that* object – the one appearing to be *there* – about which I am mistaken. To use perceived location as a means of reference, then, does not require a descriptive name. In order to point to an object, I need not represent *which* object I point to! I simply need to specify where it is for-me.[13] To do this is to *orient* to the world in a certain way, the way characteristic of first-person experience.

Thus, certain simple kinds of perceptual demonstratives are non-descriptive, identification free, and immune to error through misidentification in virtue of their position within a globally first-person form of experience, which is structured by the subject's dedicated perceptual systems. This is why joint attention is a particularly important phenomenon that, along with pair bonding, stands at the very origin of intersubjectivity. In joint attention, two subjective perspectives orient to the *same* "there," which allows for the "fusing of horizons" or a shared perceptual world. There are several varieties of joint attention, ranging from pack hunting to the human schoolroom, and most of it is beyond the scope of this essay. But because certain kinds of joint attention are vanishingly rare in non-human species, it is currently the focus of intense research among primate psychologists.[14]

Identification-free self-reference requires an empirical explanation. Subjectivity is a function of the first-person perspective. The problem of subjectivity is to explain how an identification-free first-person perspective arises. If it can be shown how the animal instantiates this perspective, then this would constitute progress in explaining what it is for experience to be "for-me." And in principle at least, it seems clear that such an explanation *can* be given. The reason we are immune to error in this way is that we possess no identifying knowledge in the first place. Surely this sort of epistemic "privilege" is not beyond the ken of naturalistic philosophy of mind. In fact explaining it won't even require a robust theory of conceptual content! I pause a moment to consider this last thought.

1.2 For-me experience vs. I-thoughts

Before moving on, a note about the terminology of "identification-free self-reference." Evans developed this terminology for a related but different purpose, in his account of *I-thoughts*. I-thoughts are thoughts in which the subject explicitly refers to herself with the pronoun "I". Evans argued that some of these I-thoughts are identification free. But the class of I-thoughts is certainly not coextensive with the class of for-me experiences. I-thoughts require something more, namely, a conceptual mastery of the first-person pronoun. Notoriously, Evans held that genuine I-thoughts must conform to the so-called *Generality Constraint*, a condition that obtains for all conceptual thinking. Evans claimed that "...if a subject can be credited with the thought that *a* is *F*, then he must have the conceptual resources for entertaining the thought that *a* is *G*, for every property *G* of which he has a conception" (Evans, 1982, p. 104). So, in order for any given thought to count as a properly conceptual thought, its thinker must also be capable of entertaining all syntactically permissible permutations of the constituents of the thought (Carruthers, 2009). Each conceptual thought must appear within a larger system of related concepts, governed by rules of composition and relations of natural deduction (Fodor, 1978). In the case of I-thoughts, this means that the thinker of "I see red" must also be able to token and understand not only, e.g., "I see green" but also "you see red" and "you see green."

As emphasized at the outset, the account of subjectivity offered here is not concerned with self-ascription in Evans' sense.[15] I am using Wittgenstein's distinction between the two uses of "I", and Shoemaker's subsequent definition of the class of P* predicates, as a way of picking out those mental states which are for-me but which are not necessarily represented or conceptualized *as* thoughts that I have. That is, for-me subjective states need not count as I-thoughts because they do not entail self-ascription. Evans was working within Russell's framework, attempting to fit identification-free self-reference into Russell's account of singular terms.[16] But part of Wittgenstein's original point, it seems to me, was that the subject-use of "I" presents a counterexample to the theory of descriptions precisely because it does not behave like a singular term. Anscombe (1975) also underscored the point that the subject use of "I" is, as it were, *subjectless*.[17] That's right – subjectivity is subjectless. The for-me-ness of experience does not respect the distinction between predication and the subject of predication, i.e., between the property and the individual who has the property. Although subjective expression in language has a perfectly grammatical surface structure, the underlying experience is not expressible as a well-formed formula within the predicate calculus, and hence violates the Generality Constraint.

Now, it might be thought that this is too quick, and that the relevant conceptual claim is the existential one, that *there is someone* who is the subject of predication, i.e., $(\exists x)Px$. This avoids giving a definite description because it contains no name. On this suggestion, the P* expressions are simply using variables instead of singular terms. The content is then *anonymous* but still perfectly "objective" or third-person, and still comfortably conceptual. Does this count as "identification free"?

No. In identification-free self-reference, if the P* claim is false, then so is the corresponding existential claim: ~P* →~(∃x)Px. If it is not I who has this headache, then no one does. But as we have seen, a false singular claim Pa certainly does not carry this implication. i.e., ~[~Pa →~(∃x)Px]. Thus, the identification-free self-ascription "I have a headache" does not mean "there is an anonymous someone who has a headache," because this carries a different implication. In particular, it would render "I do not have a headache" as "There is someone who does not have a headache," which is not the same at all. (∃x)~Px does not entail ~(∃x)Px. The * function is not just a variable that replaces a singular term. It is a place-holder for a designation that has no atomic expression and so need not meet the Generality Constraint. But this does not prevent it from expressing a truth about a for-me, P*, subjective and first-person perspective.

With that in mind, I admit that my use of the terminology of "identification-free self-reference" is somewhat awkward. Strictly speaking, the basic form of the first-person perspective is not *self*-referential, it is just the variety of referential thought picked out by the class of P* predicates. But Evans' phrase is useful to my project even so. It keeps the identification-free nature of for-me experience firmly in view, disambiguating it from the Russellian kind of immunity to error through misidentification. It emphasizes that the explanandum for the account of subjectivity is not a named entity; it is not to be analyzed through an identifying description.

There is deep disagreement about virtually every aspect of this topic. The Generality Constraint is one influential way of talking about the *Space of Reasons*, the domain of conceptual knowledge (McDowell, 1998, Sellars, 1953). At this stage the issue becomes, does for-me subjectivity entail a Space of Reasons? Many have weighed in on this question, and I'll not attempt a complete review of the literature here.[18] McDowell (1998) distinguishes between genuine Worldly Experience and mere coping, holding that a Space of Reasons is essential to the former but not the latter. There is some truth in this, but it is simpler to distinguish between I-thoughts and for-me structure. Even in the absence of I-thoughts there can be something it-is-like-for-me, and the first-person perspective originates and develops through this kind of engaged coping (see Chapters 6 and 7). Whether first-order subjective coping should be considered properly conceptual is a question which I will not attempt to decide. But note that Carruthers (2009) argues for the surprising conclusion that even invertebrates such as honey bees can meet the Generality Constraint. This suggests that the distinction between the Space of Reasons and the world with which we cope may not be as absolute as it seems. Merleau-Ponty (1962) may have been right that embodied subjectivity involves a meaning of "meaning" that has both conceptual and non-conceptual aspects.

In any event, if the for-me aspect of experience were *not* distinguishable from Evansian I-thought, then explaining it *would* require a complete naturalization of rationality itself. This would be an unreasonable burden, and it is not my goal. I take it that the identification-free form of the first-person perspective can be explained apart from the kind of robust and rational reflexivity required for self-ascription. The "unity" or coherence of experience is a loosely integrated dynamic gestalt, not a well-defined conceptual system. I next briefly characterize this holistic nature of the first-person perspective.

1.3 Situated subjectivity

I'll return to this topic more fully in Chapter 8, so my comments in this section are brief. So far, it is established that what makes an experience "for-me" is that it architecturally specifies me in virtue of certain dedicated perceptual systems. In effect I have argued that, in the P* notation introduced by Shoemaker, the * signifies the *for-me* part of what-it-is-like-for-me. This cannot be detached from what-it-is-like, as if it were a name or a variable labeling some separate entity or expression. Rather, the * accompanies P as its first-person mode of presentation. This indicates that P is experienced as *situated* relative to some perspective. That is, the * is a placeholder for the "architecture" of subjective experience. The placeholder is a blank piece of notation, but the perspective is highly differentiated. Particular pointers are situated in the stream of consciousness in relation to the ongoing flow. For example, the experience of a coffee cup is nested in an intelligible way within a larger dynamic gestalt that specifies a subjective horizon. At least, it is nested in a way that *seems* intelligible. As will eventually become clear, the relation to the larger structure need not actually *be* particularly intelligible because the total experience is not actually as unified as it may seem at first glance. Nevertheless, "the coffee cup experience" includes the first-person perspective *as a whole*.[19] This shows that casual ways of speaking about "the" experience of the coffee cup are misleading. It is an abstraction to talk of "an experience," the content of which is given as an intentional object. "The" experience exists only as part of a global subjectivity that is not determined by the intentional object alone. That is, there is no for-me experience that stands alone and has exactly the content P, where P is the predicate component of the P* judgment.

So, experience at a given time is holistically structured in a way that precludes the individuation of component "experiences," each of which exists on its own and carries an intelligible content. Michael Tye (2005) maintains a similar position in *Consciousness and Persons*:

> ...there really are no such entities as purely visual experiences or purely auditory experiences or purely olfactory experiences in normal, everyday consciousness. Where there is phenomenological unity across sense modalities, sense-specific experiences do not exist. They are figments of philosophers' and psychologists' imaginations. And there is no problem, thus, of unifying these experiences. There are no experiences to be unified. Likewise within each sense: There are not many simultaneous visual experiences, for example, combined together to form a complex visual experience. There is a single mulitmodal experience, describable in more or less rich ways. (Tye, 2005, p28)

So, the idea that various "experiences" find their place only within a single ongoing and highly differentiated structure is widely appreciated.[20] In addition, there is no point of observation *onto* experience; the stream of consciousness does flow *by* or *past* a stationary vantage point called "the first-person perspective." We're not sitting on the banks, watching the river flow.[21] Such an external vantage point would only yield a detached, third-person relation to the contents of consciousness. But there is no bird's-eye view on first-person experience.[22] Certainly, first-person experience is "gappy" (and of course there are breakdowns). But things more or less *hang together for-me* (see the Postscript). Through its perceptual-motor engagement, each animal's subjective experience assumes a minimal and situated sort unity.[23]

1.4 Objections

I'll consider two objections to the analysis of subjectivity so far advanced. The first objection concerns a well-known counterexample to the claim of immunity to error through misidentification. The second objection is more fundamental to the whole project, concerning the premise that subjectivity is analyzable in terms of the first-person perspective. I consider these objections in turn.

The initial objection is to the idea that first-hand experience is immune to error through misidentification in virtue of the dedicated architecture of embodiment (sections 1.1 & 1.2, above). As noted, information gleaned in this way concerns *contingent* relational properties about the world "from here." This contingency in embodied architecture entails that the relational properties could be otherwise. But immunity to error seems to require that they *couldn't* be otherwise, at least with respect to the question "who?" Isn't this a problem? To illustrate, simply imagine the world slightly different. Imagine that, unbeknownst to me, I am "hooked up" to someone else's perceptual systems such that the information dedicated to that other person also comes to me. Thus, there is a first-person perspective in which I have a headache, or my legs are crossed. But that perspective is actually *for* another, not for-me. In thinking that my legs are crossed, I misidentify the person whose legs are crossed as "me." Someone has a headache, but I am not that person. Doesn't this show that identification-free self-reference is not immune to error through misidentification?

I'll quickly review three prominent responses in the literature, associated with those who have most directly influenced my discussion so far: Evans, Shoemaker, and Campbell. Evans' general line is that the counterexample doesn't work because in order to be a source of knowledge (and of potential error), information systems must be *reliable*.[24] But in the counterexample, the perceptual system cannot play this reliable role in justification because the information system is short-circuited. Thus, the ensuing I-thought described in this case is not an epistemic state *at all*. Immunity to error through misidentification remains intact, since it is still the case that *either* the claim is about me (in the normal case) or (in the counterexample) it fails to be a claim about anyone at all, and the corresponding existential generalization remains false.

Unlike Evans, Shoemaker's response is not epistemic but conceptual. He holds it conceptually impossible that there be a first-person experience that is not for-me, but rather for someone else.[25] In the headache version of the counterexample, a Shoemakerian will hold fast to the intuitive position that if I apprehend a headache from the first-person perspective then I do indeed have a headache. This is true from within my first-person perspective regardless of *which cranium* is the underlying physiological cause of the pain. As for the crossed-legs case, to the extent that this case is correctly characterized as a P* experience, it still directly and internally hooks onto oneself. But again, this self-knowledge concerns only the first-person seemings, not *which legs* the experience is attached to. Correctly characterized, the case is one in which I have an experience as-of-having-my-legs-crossed, and "my" in this experience still refers to the one having the experience, not the legs. Hence, the error is still not an error of misidentification.

Finally, Campbell gives an elegant response which treats the self-ascription as a speech act: there is a simple rule governing the use of the first-person pronoun, according to which all utterances of "I" refer to the speaker. Thus, mastery of this practical rule is what guarantees the referent of the I-thought. This renders the *source* of the claim in the self-ascription irrelevant, whether descriptive, demonstrative, or experiential. For Campbell, the identification-free nature of dedicated perceptual systems remains intact in the counterexample, but this is not what grounds the immunity of the subsequent I-thoughts. The token-reflexive rule does that.

Notice something that the counterexample and all three responses have in common: They are all ultimately concerned with I-thoughts, not with the for-me structure of subjectivity. And as I see it, the really striking thing is that each account ultimately offers up a different referent of the I-thought! Evans says that the "I" is a rational knower, conceptually constituted. Shoemaker says the "I" is a neo-Lockean non-branching stream of experience. Campbell says the "I" is the concrete speaker/language user. But as far as I am concerned, the for-me structure of experience simply *leaves open* the nature of the "I" or self (cf. Anscombe, 1975). Thus, how to respond to the counterexample depends on what your theory of self is. But I have no theory of self and I am not seeking one. The question in view is rather, wherein lies the for-me nature of experience? The answer is that it lies in the identification free, dedicated structure of dynamic perceptual activity. This for-me-ness specifies a *subject position* not a self, person, or ego.[26] The counterexample is an attempt to drive a wedge between subjectivity and self, but I am happy to allow that move, because for-me experience simply does not need a self. Experience takes place *as if* there is a self or person who is always absent, always on the horizon.[27] Returning to the putative counterexample: although there is no problem in abstractly imagining one *person* hooked up to another person's stream of consciousness, there is no way to imagine one *first-person perspective* hooked up to another stream of consciousness. The former is just an internal feature of the latter. While two people can conceivably share one set of perceptual systems, this just means that two exactly similar perspectives will be specified. So the counterexample doesn't threaten my use of identification-free self-reference. I further pursue the implications of this distinction between subjectivity and self later in this chapter.

The second objection is more basic. It says that "perspective" is not what for-me-ness is about, at all. According to this objection, talk of the "first-person perspective" is strictly metaphorical; it doesn't mean anything spatiotemporal. We can investigate the perspectival form all we want, but it will always be an investigation of some particular structure *within* experience, some way experience is. This will never get at the for-me-ness, which does not vary across cases or types. For example, schizophrenic or decentered or deconstructed subjectivity is exactly as *subjective* as any other. According to this line of reasoning, consciousness is prior to all its determinations. Rather than being a "perspective" in any straightforward sense, then, perhaps subjectivity must be analyzed as an intrinsic, non-relational *inner awareness*. This influential view is found in Kriegel (2009) and Levine (2001), and it will be the focus of extended discussion in Chapters 3 and 4. A similar view is found in the phenomenological tradition, which emphasizes the way experience is *given* or self-intimating (Zahavi, 2014, 2005, 1999). Here I pause only long enough to motivate the idea that "perspective talk" in the analysis of subjectivity is not a mere metaphor, and that explaining the first-person perspective will be an explanation of the for-me aspect of experience.

The most immediate point is that there is no counterexample in which you have for-me-ness *without* perspective. This inner awareness view tends to obscure the larger field of experience (the perspective) within which "inner quality" is situated. To think of an isolated quality is to think of a subpersonal sense datum. But such a thing would *not* be for-me or subjective because every figure appears against a ground. For-me-ness attaches to the figure-ground gestalt, not to the bit within it. True, the first-person is not *merely* spatial, but talk of the perspective is not merely metaphorical, either. In fact perspective-talk is metonymical, not metaphorical; it takes an attribute for the thing meant. Subjectivity is spatial and it is temporal and it is affective. In the first-person perspective, feelings get spatiotemporally articulated. In the simplest cases, the "ground" can be something as minimal as an existential mood or a background hum. But it is enough to constitute a perspective. Ultimately, the bodily context lends perspectival structure to subjectivity.[28]

The objection also states that the investigation of perspective will always concern some particular structure *within* experience, some particular way experience is. Such an approach, then, could never explain subjectivity per se, or why there should be for-me-ness in the first place. This complaint shows that the objector may be thinking of subjectivity in transcendental terms, as providing the metaphysical conditions for consciousness.[29] At a minimum, this version of the objection makes the first-person perspective inexplicable apart from qualia themselves, and probably apart from rationality, too. But while I agree that there are serious difficulties for naturalization of both qualia and rationality, I disagree that these problems cannot be distinguished from the problem of subjectivity (see Chapter 4). In Part Two I'll indicate what the explanation of subjectivity might look like, considered apart from other problems about mind and consciousness. In short, evolutionary developmental biology will provide the general framework for explaining both the similarity and difference in the diversity of first-person character states found across vertebrate species.

1.5 Resumé: subjectivity considered as the first-person perspective

A framework for investigating subjectivity is suggested, taking the first-person perspective as the key characterization. The first-person perspective has two salient features: (A) an identification-free self-specification, and (B) a holistic and dynamic scheme that hangs together for me. A set of empirical models that could handle both A&B would constitute the beginning of an explanation of *baseline subjectivity*. It will be the task of Chapters 6 and 7 to make the case that both of these desiderata can be met with tools available in evolutionary developmental biology, affective neuroscience, and a "modular embodiment" approach to subjective cognition. In the remainder of this chapter I sharpen my analysis of the first-person by contrast with the popular narrative account of "the subject."

2 Subjectivity and the limits of narrative

Chapter 4 will be dedicated to dispelling the widely accepted idea that subjectivity presents a paradox that has no solution. But many writers charge straight ahead with various cognitive or hermeneutic theories of subjectivity and "the subject." One powerful contemporary proposal is that the first-person point of view is the point of view of a narrative self. Sometimes "the self" is spoken of as "the subject," which suggests that the first-person perspective may be constituted in the same way as the self. Narrative models of the self have proven valuable for contemporary cognitive, developmental, and moral psychology. Our self-understanding often takes a narrative form, and self-narrative can influence action and experience in subtle ways. On some versions of narrative theory (e.g., Teichert 2004, Atkins 2004), narrative is held fundamental not only to self-understanding but to the phenomenology of the first-person perspective, too. I call this approach the *narrative subject model*. In the remainder of this chapter I argue that the narrative model cannot account for subjectivity, and that the narrative self should be distinguished from the "I" of the first-person perspective. Roughly, this is because first-person narratives employ the object-use of first-person pronoun "I", identifying some person, while the distinctive features of subjectivity are marked by the non-identifying subject-use of "I" described above. The next section below describes the *narrative subject model*. I then argue that the narrative subject model does not meet the criterion of immunity to error through misidentification. The third and final section provides further evidence that the narrative self, whatever else it may be, is not a fundamental element of either situated activity or from-the-inside experience.

2.1 The narrative subject model

According to Teichert (2004) narrative offers the resources to "widen" the notion of personhood to include the self as subject. The narrative model "seizes" the first-person point of view in a way that other accounts of self and identity do not, and Teichert believes that "the gain is considerable."[30] Similarly, Atkins (2004) argues that the uniquely powerful feature of the narrative model is that first-person narratives are tied to embodied subjectivity in a special way, and that the very form of the first-person perspective is "narrative continuity."[31] She concludes that "the superiority of the narrative model arises from its presupposition of the self as embodied consciousness."[32] So the narrative subject model proposes that subjectivity is part and parcel of the narrative self. More precisely, the model makes two claims: (1) That the first-person perspective is a narrative perspective, and (2) that the narrative model unites the notions of self and subjectivity.

It is useful to distinguish at the outset between *strong narrativism* and *weak narrativism*.[33] *Strong narrativism* holds that the self is constituted through narrative, that narrativity is necessary for practical agency and for the phenomenology of agency, and hence for the self as subject.[34] By contrast, *weak narrativism* holds only that narrative self-representations are explanatory for some actions, and that although people often generate self-narratives we are not constituted by them. Weak narrativism is certainly true but incapable of telling us much about the nature of subjectivity. But the former, strong variety holds that narrative practice constitutes a real being called the "self" and that this being is the subject of first-person experience.[35] I call this the *narrative subject model*. The model has two primary components: a theory of narrative and an account what a subject is. I consider these in turn.

Narratives

A narrative is a representational structure in which past, present, and future events have a meaning that is determined by the other elements in the story.[36] Proponents of narrative begin from the Aristotelian theory of narrative as an imitation of an action. Narrative involves the imaginative ordering of diverse elements in a structure that has a beginning, middle, and end.[37] The central feature of narrative is that there is a *plot*, a connection between the elements represented. Danto (1965) contrasted this Aristotelian definition of narrative with the notion of a *chronicle*. Chronicles are lists of temporally indexed descriptions of events. Consider the following:

   1905: June 21. Birth of Jean-Paul-Charles-Aymard Sartre in Paris.
   1929: Meets Simone de Beauvoir.
   1941: Obtains freedom from prisoner-of-war camp by posing as a civilian.
   1963: Publication of *Le Mots*.
   1964: Refuses the Nobel Prize.

Since it is temporally sequenced, a chronicle is not a mere list. But it is something less than a narrative. A narrative is unified by a theme and organized with respect to its conclusion. Chronicles are merely one damned thing after another, with no intelligible connection between them.[38] By contrast, the narrative serves an explanatory function for the events described.[39] Each moment narrated finds its place in the overall structure and is rendered intelligible through its role in the plot. Roughly, a chronicle is just a bunch of stuff that happened, while a narrative represents a single intelligible thing that happened. In the narrative model of the self, then, this unitary thing is the life of a person. Atkins (2000) writes: "The continuity of my life and my identity involves more than a chronology: it also involves a coherent integration of my past, present and future in such a way that I can tolerably live."[40]

Agents make sense of their actions through their life-story. Life takes place within a web of meaning connecting past and future, and this is called *emplotment*. On the strong version of the theory, narrative does more than help people make sense of things. Emplotment is what *makes* the narrative into a story about a single being. Intuitively, one might have thought that the connection between the events in a narrative is that they are all about the same thing. For example, the chronicle above is a list of events in the life of a single person, who re-appears at each stage, and tells his own story of these events in *Le Mots*. But strong narrativism holds that the narrator appears only in and through narrative. Thus, the distinctive structural features of narrative constitute a new entity called "the subject." I now turn to a discussion of this narrative subject.

Selfhood as ipse identity

The key to the narrative subject model is the notion of *ipse identity* or *selfhood*.[41] The central question regarding ipseity is "Who did this?" and a first-person narrative is necessary to answer this question: "I did it." Unlike traditional approaches to personal identity, which focus on re-identification of an individual over time, *ipse identity* emphasizes the practical and phenomenological elements of selfhood that ground authority and responsibility. Teichert writes that "personal identity is more than a question of re-identification.... identity as selfhood has to be approached from a first-person perspective... To possess identity as selfhood means to be the subject of dynamic experience."[42] This first-person perspective is in turn treated as first-person narrative.[43] According to the narrative subject model, first-person narrative constitutes a subjective phenomenology above and beyond any objective identification, and re-identification across time cannot account for subjective *ipse* selfhood. So, according to the model first-person narrative is a non-identifying form of self-reference.

Why is the first-person *perspective* thought to be a first-person *narrative*? Subjective experience, it is argued, consists in past, present, and future, which is an expression of the structure of narrative emplotment described above. What distinguishes subjective *ipseity* from a mere chronicle of objective temporal moments is its narrative logic. Thus, according to the narrative subject model, what is distinctive about subjectivity is that it is a narrative in which "I" am the narrator.

Atkins cautions that narrativity is not simply the story of conscious experience. Nevertheless, narrative must underlie subjectivity on her view, since consciousness is internally unified only through the temporality of emplotment.[44] To clarify, this is not a *reductive* view, since it does not explain subjectivity in causal terms. Quite the opposite. The idea is that narrative inflates or bootstraps to *constitute* a real, unified subjective being that is the center of narrative experience.[45] Teichert clarifies that the goal of narrativism is not just to formulate a new criterion for personal identity, but to widen the notion of self to account the first-person, subjective point of view.[46] On the narrative subject model, to be a subject is to narrate.

If correct, the view articulated by Atkins and Teichert has important implications for the metaphysics of subjectivity. But narrative cannot do all the work required of it in this subject model. In the next section I show that narrative does not possess the criterial feature of the phenomenology of the first-person. Instead, the narrative "I" is a variety of mundane, descriptive, and (re-)identifying self-reference.

2.2 The immunity criterion

This section contains two brief parts. The first part argues that narrative cannot account for subjectivity because it is not an identification-free form of self-reference. The second part considers an objection to identification-free self-reference as the mark of subjectivity, and discusses an alternative construal of the relation between the narrative "I" and the phenomenology of the first-person.

The class of P* predicates corresponds to the psychological states apprehended from-the-inside. In the debate about consciousness, talk of the "first-person perspective" refers to the *subjectivity* of a mental state, not to *which person* the thought is properly ascribed. So, immunity to error through misidentification can serve as the mark of subjectivity. If the narrative self-subject view is to provide an adequate model of subjectivity, narrative *ipse* selfhood must exhibit immunity to error through misidentification. Recall, too, that Teichert argued that *ipse* selfhood is properly subjective in just this way: *ipseity* not a matter of re-identification, and this is what distinguishes it from "objective" personal identity. But while he claims that *ipse* selfhood is not a matter of re-identification, he also argues that it *is* a matter of there being a first-person narrative that answers the question "who?" This is an identification judgment. I argue that because first-person narratives make identifying judgments about authorship and responsibility, the narrative "I" does not meet the criterion for the phenomenology of the first-person. The properly subjective use of "I" is fundamentally *non*-narrative insofar as it lacks an identification component.

Discussion so far has shown that the mere occurrence of the pronoun "I" in a statement or narrative does not make the statement a first-person one in the relevant P* or subjective sense. In its object-use, the pronoun "I" is like any other pronoun or proper name. It does not pick out any distinctively subjective content. This is shown by the fact that we can be radically wrong about *who* we are, that is, about the identity of the person having these P* experiences. The question, to *which person* this subjective experience belongs, cannot be answered through the subjective experience itself. Knowing that these P* experiences are mine is not helpful in discovering my identity or in answering the question of "who" I am.

The narrative "I" and immunity to error

In English, one rule of thumb for determining whether a given use of "I" is an object-use is just to try translating the pronoun into a direct object referring to "me." If the resulting expression avoids redundancy, you have the object-use. For example, the meaning of "I am on the giant video screen" can quite naturally be expressed in something like the following way: "Look at that guy on the big screen, *the one with the hat – it's me!*" But when you try waving and the fellow in the picture doesn't wave, you realize someone else has the same hat. By contrast, the expression: *"The one with this pain – it's me!"* is oddly redundant because knowledge of this pain already includes the knowledge that it is mine. The wearer of the hat is not necessarily me. The subject of this pain, however, is necessarily me.

Consider the narrative confession "He ruined my life and he laughed in my face. He humiliated me, and that's why...I shot J.R." Using the translation technique just introduced, the authorial statement can be rendered: "The one who shot J.R. was me." This is an empirical claim to be evaluated independently of "I am having a memory experience as-of-shooting-J.R." The latter may certainly be taken as forensic *evidence* of the former, but that judgment would hardly be immune to error. I should emphasize that first-person mistakes of this sort are not just conceptual possibilities revealed by thought experiment or linguistic intuition. Nor are they limited to rare pathological cases. There is a growing literature on false confession in the field of forensic psychology. Kassin and colleagues have found that it is relatively simple to elicit sincere but erroneous confessions of responsibility, complete with falsely constructed first-person memories.[47] I'll quickly rehearse one such experimental design. Subjects are given a dummy computer task, say, counting words with positive or negative connotation. While being trained on the task, they are told that the computers are also being used for another ongoing job, one that is very expensive. Subjects are sternly told not push the red button, or the whole system will crash and both money and important data will be lost. As they work on the dummy task, of course, the computer shuts down. The experimenter acts very upset, and demands to know who pushed the red button. At a surprisingly high rate, subjects confess, full of shame and contrition! They begin to recount how it happened.

To return to the current example, "He ruined my life and so I shot J.R." is a first-person narrative expressing an authorial claim for an action, complete with ethical responsibility. But it is no more subjective (or embodied for that matter) than the story in which she, he, you, they, or it shot J.R. All of those narratives also make authorial claims. They only disagree about *which person* was the author. All of them, including the original first-person narrative, can misidentify the shooter. Further, the narrative is not tied to the structure of experience in any special way. Even if the story's identifying judgment is true, a zombie or robot shooter can make the judgment and tell the tale. To use Shoemaker's terminology, narrative predicates are not P* predicates. Narrative occurrences of the term "I" are object-uses of that pronoun.[48]

Now consider a second example: "I am having this memory-experience as-of previously repressed childhood abuse." This relates a from-the-inside, for-me subjective experience. As such, it expresses a P* predicate. "I" cannot fail to be the one having the memory-experience. But the content of the memory, and the resulting narrative based on it, may be false in two different ways. First, it may be false that there was abuse – perhaps I am confabulating. But it also may be false that it was *I* who was abused. Perhaps it was my sibling or friend, and I misconstrue the otherwise accurate memories of fear and pain. In that case, the narrative "I" is a mediated descriptive referent that is possibly different from the subject of the current memory-experience. Using the memory to ground an identification judgment can result in a narrative that goes something like this: "The one I now remember being abused was me." In order to assess the accuracy of the identification component of the narrative pronoun, we have to check.[49]

Further, the subject of an experience may not know that it properly belongs in the self-narrative, *despite having (P*) memories of it*. Consider a third example: "She was raped. Thank goodness it did not happen to me, I could not stand the pain." In this case, the experience itself was not forgotten or repressed. It has been remembered correctly but placed in a separate narrative that is identified as being about someone else: "The one I now remember being abused was her, not me." But this judgment about which person was the subject of the experience is certainly not guaranteed. The pain was truly experienced (was P* then) and the memory is still present (is P* now), but the narrative "I" *fails* to connect the self with either the experience or the memory. This failure of emplotment does not change the facts about the subjective experience. Clearly, then, this is not a case of identification-free self-reference. The narrative "I" is mediated (and distorted) by other factors.

Rovane's Alternative: there is no identification-free self-reference

One interesting response by the proponent of strong narrativism would be to deny the distinction between the subject-use & the object-use of "I", and to argue that *all* uses of "I" involve an identification judgment. Using an adapted version of the model of descriptive names introduced earlier, Rovane (1993) argues along just these lines. She points out that the absence of error in the P* cases does not entail that there is no identification component to the judgment. It will be sufficient if there simply are no errors. If the identification judgments are made in a highly reliable way, it may appear that that they are identification-free, when in fact they are only error free:

> Such immunity is compatible with the idea that self-reference is mediated by a description of "thinker of this thought" variety, so long as we suppose that it is accompanied by extraordinarily *reliable* self-identifying beliefs. (Rovane, 1993, p. 91)

Rovane offers a variant of the idea that "I" should be treated as a definite description, the reference of which is fixed by: "the series of psychologically related intentional episodes to which this one belongs."[50] According to Rovane, even the judgment that "I am the subject of this (P*) experience" is mediated by an identifying description after all. Thus, the proposal is fundamentally opposed to the notion of identification-free self-reference. My reading is that her alternative is indeed coherent, and that it provides the best going account of the narrative *self*, but that it is not a theory of subjectivity at all.

Rovane's version of narrativism handles objections well. For example, Christofidou (1995) objects to Rovane's view that "I" means "the series of psychologically connected intentional states to which this one belongs" on the grounds that no matter how reliable the identifying judgment, it will not be literally *immune* to error.[51] But this, I take it, is precisely the point of Rovane's account. She is trying to show that narrative self-reference need not meet any criterion of immunity, because no form of self-reference meets this criterion. She recognizes that the narrative "I" is an identifying self-concept that employs "I" in its object-use. As a proponent of narrative self-constitution, she concludes that there is no such thing as identification-free self-reference and that it is always appropriate to ask: "was it I?"[52]

Christofidou (1995) argues the diametrically opposed position that *all* uses of "I" are immune to error through misidentification. He argues that direct, identification-free self-reference is a condition of the possibility of other, properly identifying judgments about the self: "I cannot cast any light on first-person thoughts and actions by having first to establish that my thoughts and actions are connected before I ascribe them to myself. The direction should be reversed. The relation between my experience and myself, as Immanuel Kant would say, is presupposed in any "I-thoughts."[53] Insofar as Christofidou's Kantian argument is limited to a bare "I-think", rather than to *all* uses of the first-person pronoun, it is surely correct. But when the term "I" is used in natural language to refer to a public object rather than to a subjective correlate of experience, it can fail to refer in just the same way that other pronouns can fail: It can misidentify. Here is where Rovane's argument is highly relevant to the distinction between narrative self and conscious subject. Like persons and bodies, the narrative self is an object thought about via a description.

So Rovane's attempt to eliminate identification-free self-reference is well-motivated from within narrative theory. Since the narrative "I" is defeasible she must show that defeasibility is no drawback. One nice aspect of her account is that it renders certain cases of psychopathological breakdown in self-reference into rare examples in which the normally reliable, phenomenologically seamless mechanism that mediates self-reference is suddenly revealed. To illustrate this point, Rovane discusses the example of anterograde amnesia in which a subject lives in a kind of perpetual present, remembering only the past few minutes.[54] All other events since the injury are lost, though events prior to it are retained. To cope with the situation, this person keeps a diary and continually relies on it to create a self-narrative. In this case, the relation to the narrative "I" is mediated by the external notebook. Rovane argues that this is irrelevant. The amnesiac should not be denied a first-person relation to his own past simply because the mechanism of reference is mediated. This point is highly intuitive. The amnesia does not fundamentally alter the narrative relation or the self-concept based on it. It only requires an unusual mechanism for making the identification judgment. One striking feature of the case is that the atypical memory store – the notebook – is *more* reliable than normal memory.

But Rovane also recognizes the further consequence of her account. It implies that the narrative self is not the same as the embodied subject. This is the crucial difference between her version of strong narrativism and the subject model criticized in this paper. She concedes that narratively constituted selves do not coincide with the phenomenology of the subjective point of view. This allows her model to cover more kinds of self, including cases of corporate identity as well as pathological cases like the one described above. For this reason she distinguishes personal identity from human identity.[55] The former corresponds to the narrative self, while the latter is roughly the "embodied subject" spoken of by Atkins and the proponents of the narrative subject model. Thus, Rovane's claim that the amnesiac retains a first-person relation to his own past means *only* that the past person is identical to the present one, and that this identification judgment is grounded in narrative. She does not make the further claim that first-person narrative is an internal structure of subjectivity. What I am calling "Rovane's Alternative" is that all *self*-reference is mediated, but that *first-person experience is not a case of self-reference*. Or again, all uses of the term "I" are object-uses, but the term "I" does not refer to first-person experience.

Where does Rovane's Alternative leave us with regard to the narrative subject model? The situation is as follows: Either (1) there *is* a kind of self-reference that is identification free, or (2) there is *no* identification-free self-reference (Rovane's Alternative). As far as the narrative *subject* model goes, these alternatives are equivalent. If (1), then the narrative "I" does not constitute the self that is marked by identification-free uses of "I", and the narrative self lacks the criterial feature subjectivity, namely, immunity to error through misidentification. If (2), then the narrative "I" need not meet this criterion, and some version of strong narrativism might be sustained. But whatever might be sustainable will clearly be a theory of the *self*, or person not of subjectivity. So even if Rovane's rejection of identification-free self-reference is accepted, this is no help to the narrative subject theorist. In either case, the narrative self is not the subject of first-person experience.

Life without narrative

This section extends the critique of the narrative subject model by briefly discussing the basic ways in which life without narrative can and does take place. Both perceptually guided action and from-the-inside experience can occur without a narrative frame.

A strong claim on behalf of the narrative subject model is that narrative selfhood is necessary for agency. Actions have a beginning, middle, and end, exhibiting the narrative kind of unity. Recall that Aristotle defined narrative as the representation of an action. The narrative subject view radicalizes this relation, and makes narrative constitutive of action. "The strong narrativist claims (1), that there is a fundamental connection between action and narrativity: acting presupposes narrative schemata; and upon that claim, she bases a further claim, (2) that the self [i.e., first-person *ipse* selfhood] is constituted by narratives."[56] Less formally, Atkins states, "As a practical being whose existence is structured by action...who I am is structured through the textual resources of narrative."[57]

But there is a central category of action that does not require narrative selfhood.[58] The class of ecologically situated actions is corollary to the class of P* epistemic predicates identified above. Perry (2002) argues that many actions are grounded directly in *epistemic/pragmatic* relations, which he distinguishes from those guided by *permanent files* or self-concepts. Examples of epistemic/pragmatic relations include *being-at* and *in front of*.[59] Epistemic/pragmatic actions are guided by the architectural (embodied) relations between the eyes and arms, just as P* predicates are structurally connected to their subject.[60] Actively using information based on epistemic/pragmatic relations does not require that the agent also have a "file" in which the current practical situation is set (emplotted).[61]

> For many purposes we don't need notions of ourselves at all. Consider the simple act of seeing a glass of water in front of one and drinking from it. The perceptual state corresponds to a relation between an agent and a glass of water...The coordinated motion of hand, arm, and lips by which the agent gets a drink is not only normally object-in-front-of-one-effecting, but also agent-who-does-the-action-effecting. (2002, p. 208)

According to Perry, we need not keep track of the identity relation between perceiver and agent. Situated agency is identification-free in the sense that it is grounded directly in the for-me subjective structure of perception. The fact that situated agency is "plot independent" in this way does not mean that self-narrative *cannot* get connected to action, nor does it entail that narrative has no place in a theory of ethical responsibility *for* actions.[62] But ecologically situated action is important precisely because it is independent of self-conceptualization.[63] Situated action can and does take place in the absence of a narrative self.

If all this is on the right track, the narrative "I" is no index of subjectivity. Nevertheless it might be maintained that experience must be narratively structured in order to be meaningful as *one's own*. Perhaps my from-the-inside experiences can only acquire meaning (i.e., become the objects of concern) through their place in my autobiography? On the narrative subject model, my subjective concern about my experiences arises in the first instance from their emplotted relation to my narrative self. In this way, Schechtman (1996) argues that only narrative can provide the rationale for our concern about the future.[64] But even practical concern can take place from-the-inside, without a unified narrative self. In this vein, Galen Strawson (2004) denies that from-the-inside memories must be understood as something that happened to me*, where me* designates the self: [65]

> My memory of falling out of a boat has an essentially from-the-inside character, visually (the water rushing up to meet me), kinaesthetically, proprioceptively, and so on. It certainly does not follow that it carries any feeling or belief that what is remembered happened to me*, to that which I now apprehend myself to be when I am apprehending myself specifically as a self... As for my practical concern for my future, which I believe to be within the normal human range (low end), it is biologically – viscerally – grounded and autonomous in such a way that I can experience it as something immediately felt even though I have no significant sense that I* will be there in the future. (2004, p. 434)

Strawson describes his own personality, with its normal but "low end" level of concern about the future and past, as *episodic*, which he contrasts with a more *diachronic* personality type. He identifies proponents of narrativity as diachronic personality types, who are somewhat obnoxiously obsessed with the unity and meaning of their own lives. Whatever one makes of this diagnosis, Strawson's distinction between visceral from-the-inside embodied experience and *that-which-is-apprehended-as-the-self* shows the possibility of a concernful life not narrated.[66] "Visceral" concern is present in the phenomenology of the first-person perspective, and as such it is independent of self-narrative. Part of the burden of a theory of subjectivity is to account for visceral concern in the sense that Strawson indicates. One goal of this essay is to give such an account independently of any theory of "self."

3 Conclusion: life goes on

One of the most robust results from the combined history of psychology, phenomenology, and philosophy of mind is that our self-understandings are woefully inadequate. Sartre, for one, was highly critical of our narrative practices, and warned against confabulation, objectification, structuralism, and bad faith. More recently, John Campbell writes: "The puzzle is to understand how it can be that you could hold on to your understanding of the first person while being radically uncertain as to which particular thing you are."[67]

Self-narrative falls short. But the fact that we almost *never* succeed in imposing a unified meaning on the totality of our own subjective lives has been of little concern to proponents of the strong versions of narrative theory. The point is not just that first-person narratives can be false. The narrative approach to ethics holds that by ascribing the action to myself, I *identify* myself with the agent, such that I can assume responsibility.[68] First-person narrative requires an identification judgment regarding an objective *someone* to whom the predicates are attached. But the properly subjective use of "I" is fundamentally *non*-narrative insofar as it lacks an identification component. It is evident, however, that the absence of narrative selfhood in the phenomenology of the first-person disrupts neither the subjective nature of from-the-inside experience nor the ecological validity of situated agency. Beyond this, it may or may not be the case that we have an ethical responsibility to reflect on the story of our lives. But whether or not we do this, life goes on.

3
Unconscious Subjectivity

Because the notion of unconscious thought plays a crucial explanatory role in cognitive neuroscience (and philosophy of mind generally), it deserves careful scrutiny. But the nature of unconscious mental states is rarely considered problematic today. Unconscious thoughts are widely understood to be information states that cause behavior. Consciousness, by contrast, *is* considered a hard problem because it exhibits features that information functions do not. Consciousness is qualitative, subjective, and strongly connected with agency. These dimensions of conscious thought are not found in the Swiss Army knife assortment of specialized computational mechanisms featured in mainstream cognitive neuroscience. Roughly, consciousness resists explanation because information can always be processed without it, as any zombie will surely tell you (Chalmers, 1996 & 1998; Levine, 2001).

But there is a paradigmatic variety of unconscious thought that has much in common with the conscious states just described. Some unconscious thoughts have subjective character. There is something it is like to be in these mental states, such that they matter to a particular subject. For us as individuals, it is not just "informative" to learn of these states, as when we learn something about anatomy. Rather, they have first-person psychological and practical relevance for individual life experience. It has been remarked that finding out about our unconscious motives is not just an intriguing intellectual exercise but a moral obligation (Rorty, 1991, p. 145). I set up the discussion of unconscious subjectivity by returning briefly to the notion of for-me mental states.

1 Preliminary on for-me mental states and self-awareness

The first task is to disentangle for-me-ness and conscious awareness. According to many philosophers, this is not possible. For example,

Uriah Kriegel's *Subjective Consciousness: A Self-Representational Theory* (2009) is built on the contrary assumption that subjectivity is what makes a mental state conscious at all. According to self-representationalism, subjectivity *just is* awareness. If that is right, then there can be no such thing as unconscious subjectivity. According to that view, a mental state has subjective character when it represents itself in the right way. Roughly, this "right way" is such that the representation comes to constitute a reflexive awareness. The self-representational framework for inner awareness is based on a distinction between representations for-me and representations merely in me (Kriegel, 2009, p. 16). A good example of a representation that is in me but not for-me is the retinal image. I cannot see my retinal image. The in me/for-me distinction is also fundamental to this essay – it is the common starting point Kriegel and I share. But Kriegel cashes out the notion of for-me mental states in terms of awareness:

> As for subjective character, to say that my experience has subjective character is to point to a certain *awareness* I have of my experience. Conscious experiences are not states we may *host*, as it were, unawares. Freudian suppressed states, sub-personal states, and a variety of other unconscious states may occur within us completely unbeknownst to us, but the intuition is that conscious experiences are different. A mental state of which one is completely unaware is not a conscious experience. In this sense, my conscious experience is not only *in me*, it is also *for me*. (p. 8)

Here Kriegel fully embraces the paradoxical idea that subjectivity is a kind of inner awareness that cannot remain unknown (see Chapter 4 for more on this argument). The basic assumption is just that subjectivity is whatever it is that makes us consciously aware. According to self-representationalism, the for-me/in me distinction not only picks out all and only subjective states; a fortiori, it also picks out all and only those states we are aware of. But this may be resisted. While it is true that "a mental state of which one is completely unaware is not a conscious experience," this does not show that such a mental state cannot be for-me or subjective. Also note that in the passage quoted, "Freudian suppressed states" are lumped in with all the other varieties of unconscious cognition. But the psychoanalytic unconscious is unlike the cognitive unconscious on precisely this score – it is subjective. I take up this topic toward the end of this chapter.

Like Levine's (2001) *Purple Haze*, several of the core arguments in *Subjective Consciousness* rely on the idea that subjectivity is the "awareness-maker." For example, to reject the possibility that subjectivity is simply a nonrepresentational "intrinsic glow" or background hum in consciousness,[1] Kriegel offers the *argument from awareness-making* as follows. An intrinsic glow would be unstructured and nonrepresentational. But no unstructured and nonrepresentational process could suffice to make the subject aware of her experience. Therefore subjectivity is not merely an intrinsic glow (2009, p. 104). My take on this argument is that although it is true that intrinsic glow is not an awareness-maker, this is not a reason to reject the intrinsic glow approach because the subjectivity-maker need not also be the awareneness-maker. Also note that "awareness" is explicitly understood in epistemic terms in this argument. Kriegel rightly emphasizes that awareness is knowledge. And in my view this is precisely the reason *not* to equate subjectivity with awareness. First-person experience can exist without our knowledge (see Chapter 4). As the name of Kriegel's argument suggests, the crucial premise is that *subjective character is what makes the subject aware of her experience*. The support offered for this premise is that the alternatives are unpalatable. To reject the premise is either (a) to be radically eliminativist about subjectivity, or (b) to adopt an "intermediate" view, to the effect that we could be unaware of some subjective states:

> ...But the intermediate position, according to which conscious states are experientially given to us and possess subjective character, and yet we are completely unaware of them, is fundamentally unstable and quite possibly incoherent. (p. 105)

To clarify, the position described is said to be intermediate between the extremes of (a) eliminativism about subjectivity and (b) the immediate inner awareness view embraced by Kriegel. And with the right caveats in place, the "unstable" position described is the one I hold. We can remain unaware of subjective states. Further, I doubt whether anything is straightforwardly "given to us" experientially (see Chapter 4). But the immediate point is that, again, Kriegel's text just assumes that only conscious states can possess subjective character. So this cannot provide *support* for the claim that subjectivity is the awareness-maker (consciousness-maker). And if subjectivity is not the awareness-maker, then the way is open both for other accounts of subjectivity and for other accounts of awareness.[2]

Behind the argument from awareness-making there is a more fundamental objection to the very idea of unconscious subjectivity. According to self-representationalism, the whole reason to write on subjectivity in the first place is that it is *that which gives rise to the mystery of consciousness* (2009, pp. 1–2). If so then the term "subjectivity" should be reserved for whatever it is that makes the difference between conscious and unconscious thought, since therein lies the mystery. Obviously whatever I am calling "unconscious subjectivity" cannot be *that*. In short, the basic objection is that distinguishing between subjectivity and conscious awareness merely changes the topic.

My reply is twofold. First, there is no one mysterious property, "subjectivity," that gives rise to consciousness. Instead, there are several conditions that are individually necessary and jointly sufficient, and subjectivity is among them (other conditions include phenomenal quality and metacognitive awareness). Each of these elements can take place unconsciously. While each component is mysterious in its own right, I suggest that today subjectivity is *less* mysterious than the sheer ontology of qualia, and that this constitutes progress on the explanatory gap (see Chapters 6 and 7). Second, according to self-representationalism, subjectivity is the genus for consciousness while phenomenal quality is the specific difference. That is, being subjective is what makes a mental state conscious "at all," while bearing a particular quality is what makes a mental state the particular conscious experience it is. But the genus-species relation is imprecise. Certainly all conscious states are subjective – all *but not only* conscious states! There are more species in the genus than self-representationalism recognizes. So, distinguishing subjectivity from awareness is not changing the topic.

With all this in hand, I finally turn to making the case for unconscious subjectivity. The discussion comes in three large chunks. In part one I derive a distinction between conscious and unconscious subjectivity from a critique of Block's (2011, 2007, 1995) important but problematic distinction between phenomenal consciousness and access consciousness. Reflection on this distinction reveals Block's tacit thinking on subjectivity, which I argue is not radical enough. In part two, I rehearse the debate surrounding the so-called *mesh argument*, and suggest that the preponderance of empirical evidence points toward the actual existence of unconscious subjectivity. In part three, I contrast two influential models of unconscious thought: cognitive and psychoanalytic. The currently mainstream version of the cognitive model assumes that no unconscious mentation is subjective. But the recognition of unconscious subjectivity points toward renewed resources for understanding unconscious emotions and other first-person states of which we are unaware.

2 Unconscious Subjectivity Part I: What-it-is-like vs. awareness

The idea of a first-person mental state that is not conscious may initially appear incoherent, and many philosophers from at least the time of Locke have assumed that it is so. But it is not.[3] The concept of unconscious subjectivity offers an improvement on Block's (1995, 2011) important but flawed notion of phenomenal consciousness without access consciousness. The valuable core of Block's insight is best expressed, not as a distinction between two kinds of consciousness but as a related distinction between conscious and unconscious subjectivity. Despite Block's critique of received philosophical views about consciousness, he still tacitly assumes that the subjectivity of a mental state is both necessary and sufficient for calling it conscious. But the relation between consciousness and subjectivity is not biconditional because, as reflection on Block's own analysis indicates, conscious awareness is sufficient but *not* necessary for subjective thought.

Begin with Block's well-known claim that "consciousness" denotes a mongrel concept, an admixture of distinct ideas. He argues that the mutt consists of two breeds: phenomenal consciousness (or *p-consciousness*) and access consciousness (or *a-consciousness*).[4] For Block, p-consciousness corresponds roughly to what-it-is-like: any state of an organism such that there is something it-is-like for the organism to be in that state is p-conscious. A-consciousness corresponds roughly with known experience: any state of an organism such that the organism knows what-it-is-like is a-conscious. Block argued that each of these kinds of consciousness might occur without the other. That is, there might be both *a-only-consciousness* and *p-only-consciousness*.[5] But both of these putative breeds of "consciousness" are very odd indeed.

What might p-only consciousness amount to? Block characterized access in terms of cognition and rationality. Access consciousness brings "inferential promiscuity," because its content is "poised for use as a premise in reasoning," and "poised for the rational control of action."[6] To access an experience, then, is to *know* (represent) what experience is being accessed. So p-only-consciousness will lack the rational/epistemic element conferred by accessibility. The rational component, in turn, can be understood in terms of metacognition.[7] A belief, judgment, or feeling is *metacognitive* if it is directed at other beliefs or attitudes of the cognizer. Having rational access is metacognitive because it involves a knowledge representation about the conscious state, bringing it into coherence with other beliefs in a way that is relatively robust ("promiscuous"). Thus access consciousness is a reflexively metacognitive form of consciousness.

Locke defined consciousness as the *awareness of what passes in one's own mind*, and this definition is closely related to what Block calls access consciousness. By contrast, p-only-consciousness is simply the compresence of qualitative and subjective character.[8] This Blockian distinction nicely allows for the possibility that one can be in a p-conscious (and hence subjective) state without having access to it. Or, to put the point in Lockean terms, it allows for the possibility that one might not be aware of *all* that passes in one's own mind. An example of an unaccessed phenomenal state might be the tune in my head that I don't realize is there. I later become conscious of the fact that I have been "hearing" the song for a while. Sometimes I even wake up from sleep in the middle of a tune. And of course, there are innumerable romance stories about being in love with someone but not realizing it.

Although Block's distinction between p-consciousness and a-consciousness created an uproar in the specialized literature on the problem of consciousness (and is still contested there), he is by no means alone in philosophy and psychology more generally. For example, in a somewhat different but parallel part of the contemporary literature, Dan Haybron (2010) makes a related point. He argues that introspective judgments are highly fallible, and that as a result we are often in a state of *affective ignorance*, i.e., ignorance with respect to the character of our own experience. He concludes that cases of phenomenal consciousness without access are common in everyday life: "Our powers to assess our own happiness – and more broadly, our experience of life – are weaker and less reliable than we tend to suppose. We are...vulnerable to what I will call – for want of a better name – *affective ignorance*"(Haybron, 2010, p. 2). Haybron, like Block, needs to establish criteria for distinguishing reliable from bogus introspection. Nevertheless, it does seem that we can be mistaken about our own experiences (see Chapter 4). For example, chronically stressed individuals who cope with all sorts of pain, exhaustion, and so forth may not describe their condition as "stressed out." Haybron's term "affective ignorance" emphasizes the epistemic component to these cases, rather than the functional element emphasized by Block's term "access." Both philosophers argue that there is something it-is-like for these subjects, but this something is not known. I next argue that what Block identified as p-only consciousness is better understood as unconscious subjectivity.

2.1 From p-only consciousness to unconscious subjectivity

By Blockian lights, p-only mental states are still conscious even though they are not known by the subject. Why? There is something it-is-like to be in them. By contrast, I hold that p-only states are unconscious. Why? Because consciousness has an inescapably epistemic dimension, a dimension often picked out by the term "awareness."

I begin by adopting the straightforward definition of conscious mental states already introduced: conscious states are states we are aware of having. Call this the *neo-Lockean definition* of consciousness. Locke assumed that *all* thought is conscious. The traditional Lockean biconditional of *thought* ↔ *consciousness* is the historical antecedent of the assumed biconditional of *subjectivity* ↔ *consciousness* criticized in this chapter. But the *neo*-Lockean definition I adopt only mandates that we must be aware of all our *conscious* thoughts. This updated definition, then, is neutral with respect to whether there can be unconscious subjectivity. Thus the issue becomes empirical.

The neo-Lockean definition I adopt is shared by many contemporary researchers, including Kriegel. But it is rejected by Block, who touts p-only-consciousness as the counterexample to it. Neo-Lockeans must therefore respond to the putative counterexample, either by holding that we really are aware of p-only states in some way, or that they are unconscious after all. I take the latter course while Kriegel takes the former, attempting to embrace both a-consciousness and p-consciousness as differing forms of awareness, *focal awareness* and *peripheral awareness*, respectively. He compares p-only-consciousness to the unfocused periphery of the visual field. Objects in the periphery, he points out, are not easily reported or "accessed." But he argues that there is a sort of peripheral awareness of p-only states, and that this still counts as *some* awareness, which suffices to save the neo-Lockean definition from the counterexample: "When it is claimed that conscious states are states we are aware of, the claim is not that we are focally aware of every conscious state we are in…. The claim is rather that we are at least peripherally aware of every conscious state we are in."(ibid. pp. 18).

Kriegel employs the focal/peripheral distinction in order to find some sense of "awareness" applicable to the p-only cases that seem, intuitively, outside awareness. This move seems forced. It's apparent motivation lies only in a prior theoretical commitment to the inner awareness conception of subjectivity. In place of the focal/peripheral distinction, I adopt the more familiar distinction between explicit and implicit representation, in which unconscious thoughts are those that are represented implicitly. This fits more naturally with the key idea that p-only mental states may be *accessible but not accessed*. Gaining access to a p-only state, then, is a matter of forming an explicit representation of its previously implicit (yet already for-me) content. Consider that even the notion of peripheral awareness requires at least some *occurrent* level of access. Objects in the periphery are, after all, *there*. Since self-representationalism is the view that consciousness is equivalent to occurring inner awareness, Kriegel would surely accept this. But now notice that to say an aspect of occurrent awareness is "peripherally conscious" is not at all the same as saying that there is something it-is-like which is accessible but not accessed. The latter does not mean that there is any occurrent access or knowledge. In short, even peripheral awareness must still be *explicit*, while unaccessed p-only states can be implicit.[9]

The valuable core of Block's idea is that there can be mental states in which the subject is unaware of what-it-is-like despite there being something it-is-like. But although Block has picked out an important category with the notion of p-only-consciousness, his analysis is hampered by his rejection of the neo-Lockean idea that consciousness consists in explicit awareness. The logic of Block's position can be better understood by looking more carefully at his tacit thinking on subjectivity itself.

2.1 Mongrel consciousness and for-me mental states

Again: Block holds consciousness to be a mongrel, a cross-breed of two distinct concepts – access and phenomenal. Neither of these exactly corresponds with the neo-Lockean definition, since "awareness" intuitively includes elements of both kinds of consciousness. Above I argued that awareness is roughly allied with Block's notion of access consciousness. But strictly speaking, "access" in Block's sense must be something *less* than awareness. A-consciousness can theoretically lack phenomenal quality in a way that neo-Lockean conscious awareness cannot. This is worth considering more carefully. Can we make sense of the idea of a-only-consciousness? What must it involve?

An access-only mental state would not just be a mundane case of metacognition without phenomenal qualities. There are certainly countless such cases. But a-only-consciousness would be a case of *consciousness* without quality. Block proposed "superblindsight" as a hypothetical example. Blindsight is a much-discussed phenomenon in which subjects with blindness in part of their visual field, when prompted, can guess what stimulus was presented to the blindfield with better than random accuracy. Thus, blindsighted patients have some access to visual information even in the absence of conscious awareness. In Block's hypothetical superblindsight, the performance of these phenomenologically blind subjects would be rationally optimized. They would retain reliable access to visual information, which they could use to guide intentional behavior in the rational, knowledgeable way characteristic of access consciousness.

Grant for a moment that superblindsight could occur. The question is, what would make it a form of *consciousness*? By hypothesis, there is no phenomenal quality to superblindsight. So there must then be something else, something *other than phenomenal quality*, to distinguish these access-only states from *non*-conscious metacognitive states. If not, the appellation "consciousness" is not merely mongrel but really arbitrary. But I don't think Block is using the idea arbitrarily. Beneath their mongrel nature, there is something his two concepts of consciousness have in common after all – something *less* than inner awareness – that distinguishes them from subpersonal neurocognitive mechanisms and processes. The natural candidate here is **subjectivity**. Presumably, even access-only consciousness is accessible *for someone*. To deny this would leave Block with the dubious claim that a-only consciousness can occur, not only without phenomenology, but also without a subject. So Block assumes subjectivity to be both necessary and sufficient for consciousness of whatever type. On this interpretation, all and only for-me mental states would be conscious states. This is an unhappy result for Block, because now "consciousness" does not denote a mongrel after all! Instead it looks like Block covertly holds a unified view that is actually quite traditional. Thus, the Blockian critique of received ideas about consciousness is not radical enough. He successfully distinguishes *something* from both (a) neo-Lockean conscious awareness, and (b) mundane metacognition. But he persists in calling this something "consciousness." He does not recognize that the newly identified category is simply that of subjectivity, i.e., it is the category of for-me mentation that is necessary but not sufficient for conscious awareness. Inasmuch as p-only states are distinguished from nonconscious subpersonal processes on the basis of their subjective character, and from conscious states in virtue of our lack of awareness of them, they are best understood as unconscious subjectivity.

In calling p-only mental states "conscious," Block ironically reaffirms the identification of subjectivity and consciousness, even at the very moment he distinguishes what-it-is-like-for-me from my knowledge of it. Self-representationalism, meanwhile, makes the equation of subjectivity and conscious awareness into the centerpiece of a whole philosophy. But if the above is on the right track, the relation between subjectivity and consciousness is not biconditional. As noted, Kriegel argues for the second half of the biconditional – that subjectivity is sufficient for conscious awareness – as follows: "...a mental state of which the subject is completely unaware is a subpersonal, and therefore unconscious, state."[10] But this is precisely what is at issue. If instead subjectivity is first-person referential structure, while inner awareness is a form of metacognition, then there is no reason that a mental state cannot be both subjective and beyond my awareness. I might merely "live" the phenomenal quality or "enact" it, even in the absence of explicit awareness.[11]

So unconscious subjectivity is a live conceptual possibility, and whether it actually exists is an empirical question. Is there any empirical evidence one way or the other? An indirect case can be extracted from the ongoing debate about a problem raised by Block's idea of p-only mental states, the problem of unreportable experience.[12]

3 Unconscious Subjectivity Part II: The mesh argument and the problem of unreportable experience

Block's rejection of the neo-Lockean definition of consciousness is, in part, polemical. He employed the distinction between cognitive access and phenomenology as part of a critique of experimental methods for studying consciousness. One of the primary tools he criticized is the *contrastive method*, in which introspective awareness is treated as a variable.[13] In the contrastive method, the subject's reported awareness of a stimulus is taken as an empirical measure of its conscious status. The behavioral consequences of exposure to masked or below threshold stimuli can be compared to those of consciously perceived stimuli. In this way the phenomenon of semantic priming has been well documented. But Block argued that the contrastive method only reveals what is access conscious, and so cannot discover the facts about phenomenal consciousness. This creates *the problem of unreportable experience*: Because consciousness cannot be detected directly, cognitive science depends on behavioral reports. Subjects must be asked to introspect and report whether they are aware of the stimulus.[14] But if experience and reportability can come apart, then the contrastive method will be vulnerable to false negatives, or "misses."[15] That is, in cases where subjects give a negative behavioral report there may yet be experience.

The best response to this problem is to find other empirical evidence about unconscious for-me states. This is exactly the strategy pursued by various researchers including Block himself and Victor Lamme (2006, 2007). In this section I first review the problem of unreportable experience and the empirical response to it, dubbed the "mesh argument." Discussion will show that there is indirect empirical evidence supporting an abductive inference that there are first-person mental states occupying a middle ground between prototypically conscious states and nonconscious, subpersonal information states. Block-heads argue that these are p-only conscious states. But I argue that the best interpretation of the data is that these states are subjective but not conscious (cf. Cohen & Dennett, 2011).

Some neurospsychologists, such as Victor Lamme (2006, 2007) try to evade the problem of unreportable experience by brute force. From Lamme's point of view, the problem of unreportable experience stems from treating experience as an entirely mental or psychological concept (Lamme, 2006, p. 494). He argues for a "true neural stance on consciousness," in which the term is redefined to include a neural criterion. This would make it possible to detect "consciousness" by detecting specific kinds of neural activity, regardless of overt behavioral report. He nominates *local re-entrant processing* as the relevant neural activity, and distinguishes this local activity from (a) feedforward activity, which remains unconscious, and, (b) *global reentrance*, which is large scale neural integration necessary for access and behavioral report.[16] By adding neural criteria, he argues, p-only-mental states can be rendered safe for neuroscience. This move is not entirely unreasonable. Neurophilosophy aims at finding reflective equilibrium between philosophy, neurobiology, and psychology.[17] Accepting this means, in part, accepting that concepts like consciousness, experience, and subjectivity can be modulated and improved by contact with neurobiology. Indeed, neurophilosophers have found modest success in reworking the received concept of *representation* in light of the empirical evidence about the nature of neural processing (see Chapter 7). But of course, researchers are *not* free to simply jettison philosophical or psychological issues when they become inconvenient. If neural criteria for experience are to be adopted, there must be good psychological evidence and methodological justification.

There is indeed a body of indirect data about unreportable experience. A range of well-studied phenomena behave *more like* prototypically conscious experiences than like prototypically unconscious cognitive processes. The evidence goes back to the classic experiments on iconic memory reported by Sperling (1960) and reproduced by Landman, et al,

(2003). *Iconic memory* is thought by some to be the maintenance of information "online" long enough for other neural networks to synchronize to it. In the experiments, subjects are presented with an array of 8–12 numbers and/or letters arranged in rows and columns. The presentation lasts for a few hundred milliseconds, just at threshold duration. In the absence of a cue, subjects can afterwards correctly report approximately four items from the array. However, if cued to attend to one line or column in the presentation, their reports are nearly flawless. The cue can be to any part of the array, and subjects will report with equal success. It seems, then, that people must "see" the whole array presented even though they can only access part of it. And this iconic stage has several features in common with conscious experience, such as feature binding and figure-ground segregation. This result contrasts sharply with findings about backmasked stimuli. In backmasking paradigms, the target stimulus is similarly flashed, but then immediately replaced by an interfering or "masking" stimulus that the subject becomes aware of in its stead. Masked visual information still gets into the system, but in ways that are more typical of *un*conscious processing. Although there are priming effects (evidence of feature detection), deeply masked stimuli do not display feature binding. Importantly, cues are no use in the case of deep masking. Subjective reportability does not improve in the way it does in the iconic memory paradigm. Thus, there is evidence of a processing stage (iconic memory) that has all the prototypical features of visual awareness *except* reportability, and there is also evidence that some stimuli can get into the cognitive system without reaching this iconic stage.

There are competing interpretations of the iconic memory phenomenon. But the idea is that iconic memory shows that we can entertain more for-me information than we can access. Block (2007, 2011) calls this *phenomenal overflow,* and argues that it is evidence of p-only consciousness. Further evidence about phenomenal overflow can be gleaned from data obtained when attention is manipulated in a dual-task paradigm. People can perform surprisingly well under these conditions, and it is argued that this also indicates conscious experience preceding cognitive access. For example, people can report the *gist of a scene* very accurately, even when engaged in a distracting search task and when the scene is presented unexpectedly and for a mere 30 milliseconds. The gist does not include details and changes, but does have content. So "gist perception" does not require attention. Similarly, subjects engaged in a perceptual search task can still tell whether a quickly flashed picture contains a male or female face. Performance of this secondary identification task does not inhibit performance of the primary search task. Thus, the face reached "awareness" (iconic memory) despite not utilizing any of the resources for attention.

Block (2011, 2007) takes these and other phenomena as evidence of p-only consciousness, or "awareness" without attention. He further argues this interpretation fits well (meshes) with our best current neuropsychological models of consciousness in terms of local re-entrant processing. This "mesh argument" is an abductive inference to the best available model, making best sense of the range of available data. Though far from conclusive, it is an attractive argument. The interpretation of the iconic overflow studies relies on an assumption that any stage of neural processing that bears other prototypical features of consciousness probably also shares the feature of phenomenality. There is no truly compelling reason to force acceptance of this, though Lamme attempts to ground the inference in parsimony (2007, p. 512). Cohen and Dennett (2011), however, argue that items in iconic memory are not conscious until they are attended and thereby brought into awareness (Cohen, MA, & Dennett, DC, 2011. See also Dehaene, S et al, 2006). But if unconscious subjectivity is allowed as a third possibility, then each interpretation may be partly correct: iconic memory is for-me or first-person, but can remain unconscious. This conclusion is better supported than is Block's inference that iconic memory is conscious. The priming and cueing data indicate that there is figure ground segregation and feature binding in the iconic array (but not in the masked stimuli). These gestalt structural elements are classically perspectival – they arise as part of the form of first-person experience. A figure is always "there," oriented along a "ground" in a way that refers back to the perspective "from here." So the iconic array is *perceived*, i.e., it takes the form of embodied for-me subjectivity. But this perception is not known until attended.

3.1 Attention

Cohen is an advocate of the view that attention is the key to consciousness. This general approach has many proponents, including Prinz (2012). On this view consciousness cannot be separated from function, so the very idea of p-only consciousness is a non-starter. This attitude is highly congenial to the neo-Lockean definition of consciousness adopted in the previous section, according to which conscious awareness has an essential epistemic component connecting it with access. The advocates of attention appeal to their own range of pet phenomena to support their view. In the well-documented *inattentional blindness*, subjects fail to notice unexpected stimuli while engaged in an attention-consuming task.[18] In *change blindness*, subjects fail to detect changes to a scene that occur during a blink or photo edit. Similarly, in the *attentional blink*, subjects do not observe the second of a pair of stimuli when it is flashed 200–500 milliseconds after the first. Advocates of attention argue that all these cases are evidence that unattended features just are not consciously seen in the first place, even though information is getting into the system.[19] As above, to the extent that it can be shown that the unattended/unnoticed phenomena exhibit features of the first-person perspective, I take this as evidence of visual perception that is subjective but unconscious.

Even more recently, Cohen, et al (2012) argue that attention is necessary but not sufficient for consciousness (a position obviously similar to the one argued here, about subjectivity). To make the case, they supplement the evidence about inattentional blindness just rehearsed with two further claims: that attention is *insufficient* for awareness, and that attention *cannot be dissociated* from awareness. Both claims are contested, but the second is more controversial. The first claim, that attention is insufficient for awareness, rests primarily on the fact that directed attention can increase priming effects for masked, crowded, flash-suppressed, and sub-threshold stimuli (Shin, K. et al, 2009, Van den Bussche, E. et al, 2010). This shows that attention can affect the degree to which a stimulus is *unconsciously* processed (Cohen, Cavanagh, et al 2012, p. 3). Thus attention can take place without consciousness. I leave it to proponents of the theory that consciousness and attention are the same to respond to this evidence. But the second claim is more central to the debate about Blockian p-only consciousness. If attention and consciousness cannot be dissociated, this would be (nondemonstrative) grist for the mill that p-only consciousness is a philosophical flight of fancy, and that the mesh argument does not go through.

In denying that attention can be dissociated from awareness, Cohen, Cavanagh, et al (2012) reject Block and Lamme's interpretation of dual-task attention data (gist of a scene, etc.), recruited as part of the mesh argument. Cohen, Cavanagh, et al argue that the mesh argument is not empirically rigorous. Because the *absence* of an attention effect is a null finding, it should be treated very cautiously. In particular, each null effect should be sustained across all relevant experimental paradigms (ibid. p. 3). That is, there should be *no* manipulation of attention that affects the finding. And, they argue, this criterion is not met. For example, both the perception of the gist of a scene and detection of an unattended face can be hampered in an attentional blink paradigm. So although a given experimental paradigm may suggest the dissociability of attention and awareness, no such result withstands other relevant tests. What the data indicate is just that attention is not a monolithic resource. The authors conclude that the best interpretation of the existing data is that attention is necessary but not sufficient for consciousness, and further, that attention is a primary *cause* of consciousness. I am sympathetic to both conclusions (see Chapter 8).

What, finally, of the relation between attention and subjectivity? It appears that each is necessary but not sufficient for consciousness. Is it possible that we are talking about the same thing under different names? It seems not. First, the figure-ground and feature binding effects in iconic memory seem to take place regardless of attentional cueing. Second, there is a whole range of affective and emotional phenomena that are for-me or subjective but that need not be attended or accessed. Background moods and stressors are a case in point. There is something it-is-like to have these affections, whether or not they are attended. And like the iconic array, they can be perceived in first-person form without becoming conscious. So, they behave very much *like* consciously perceived feelings, but the subject is unaware of them (Prinz, 2005, p. 16).

In this respect, emotional consciousness may be quite comparable to visual consciousness, even though the latter is more typically "cognitive." Classically, Jackendoff (1987) proposed a three-tiered hierarchical model of vision. The first tier is the subpersonal and anonymous level of unconscious feature detection.[20] The second tier "binds" these bits together into coherent contours, and represents whole objects in perspective. In the third tier of processing, general features are abstracted and objects are represented apart from any particular point of view. Jackendoff originally maintained that the second tier corresponds to conscious awareness. Not surprisingly, I hold that although this intermediate tier might be subjective (since it has first-person structure), it need not be conscious. Prinz (2005 and elsewhere) also argues that both visual and emotional intermediate level perception can take place without conscious awareness. Like Cohen and company, he appeals to the attention data to show that perception can be unconscious. According to Prinz's AIR model, consciousness consists in (A)ttended (I)ntermediate level (R)epresentations. I will suggest below that when first-person representations are affectively engaged they come to be situated within a world of concern for the organism, and thus they become more properly subjective. For the present, I conclude that the two views are compatible but not equivalent: Attention and subjectivity are both necessary but not sufficient for consciousness.

3.2 The minimal conscious state

Before leaving the topic of unreportable experience, consider another kind of case drawn from neuroethics. There is a contemporary debate about a proposed diagnosis of *minimal conscious state* (mcs), a condition argued to be distinct from the permanent vegetative state (pvs). Among patients with severe brain injury, a continuum of neural activity can be observed. Across this continuum, globally integrated function is destroyed – the kind of re-entrant processing thought to be required for behavioral report. But in a subset of cases currently diagnosed as pvs, local "islands" of neural activity remain. Some of this local activity may fit Lamme's model of local re-entrant processing discussed above, which he proposed as the neural signature of p-only mental states. In a series of studies and articles, Nicholas Schiff and colleagues argue that these islands of activity support a diagnosis of mcs, as opposed to pvs (e,g, Giacino, Fins, Laureys & Schiff, 2014; Schiff & Fins, 2003). This diagnosis, swiftly gaining popularity, might soon become ethically and legally relevant to the decision to withdraw life-support. So, here is a real kind of case in which a proposed neurophilosophical change to the received concept of consciousness might actually gain traction. These patients would be classified as "conscious" *entirely* on neural grounds, rather than on the basis of behavioral report.

So the case of the minimal conscious state lends further plausibility to the idea of unconscious subjectivity, supplementing the mesh argument. It is possible that there is something it-is-like to be in a minimally conscious state (something not very pleasant), but it is much less plausible that the subject has knowledge *of* what-it-is-like. The latter is less plausible because knowledge requires conceptualization and sensory coherence, and these in turn require larger-scale neurocognitive integration. In this respect the minimal conscious state is unlike locked-in syndrome, in which the patient remains introspectively aware but is simply unable to generate overt behavior.

So far this chapter has been dedicated to criticizing the assumption that all subjectivity is conscious. Most cognitive neuroscientists share this assumption, but conceive it in its equivalent contrapositive form: all *un*conscious thought is *non*-subjective. The next step is to examine the relation between subjectivity and *un*conscious thought.

4 Unconscious Subjectivity Part III: The incompleteness of the cognitive unconscious

Two general models of unconscious thought may be distinguished: *cognitive* and *psychoanalytic*. Several prominent researchers in the neuroscience of emotion such as Joseph LeDoux (2002) and Antonio Damasio (1999, 2003) explicitly adopt the cognitive model of unconscious thought. Some observers have argued that the notion of unconscious thought is the primary theoretical insight that cognitive science has preserved from Freud, and further, that cognitive science is the ultimate realization of Freud's research program (Kitcher, 1992). But the cognitive approach to the unconscious is quite different from its psychoanalytic counterpart. The two models do not mark out identical sets of phenomena. Because there are unconscious emotions with for-me subjective significance, a classically cognitive or subpersonal model is not sufficient. Now that I have disentangled the notions of for-me-ness and awareness, I can utilize the for me/in me distinction to draw the contrast between these two models of unconscious thought. Representations in the cognitive unconscious are in me, but not for-me. Although Freudian metapsychology was largely unsuccessful, Freud's lasting contribution was not just the idea of unconscious thought, but the idea that *subjective* thought is often unconscious.

4.1 Two models of the unconscious

The *cognitive unconscious* is constituted by the machinery that underlies the mind.[21] Consider a few of the processes involved in conducting a conversation: distinguishing the stream of words from background noise; parsing the soundstream into phonetic and morphemic structures; applying semantic and pragmatic models to the sentence as a whole; filling in gaps in the discourse. In understanding spoken language, we perform these and other complex information functions automatically and without conscious control. Many of these processes are not merely unnoticed, they are beyond notice. Introspectively, we cannot tell how or when they take place.

The vast majority of information processing is of this variety. These subpersonal mechanisms are the special province of cognitive science. With the aid of neuroimaging and other techniques, researchers are making strides toward mapping information processes onto brain structures. Though cognitive psychologists once insisted that they were concerned only with a mid-level functional architecture independent of physiology, now the primary task for cognitive science is the identification of information processes with particular neural mechanisms. Computational functions first defined in terms of information theory are now seen as biofunctional mechanisms, underwritten by natural selection. The general goal is now to show how the ensemble of these neurocognitive mechanisms causes behavior. In short, cognitive psychology has become cognitive neuroscience.

The cognitive unconscious can be characterized as a "bottom up" approach to mind. It aims to show how experience and behavior are caused by brain processes. The cognitive unconscious is offered as a causal explanation of how there come to be conscious subjects.[22] To avoid circularity, this explanation cannot advert to subsystems of the brain that have the very subjective properties it seeks to explain (Noe, 2006, p. 21; Dennett, 1975, pp. 170–171). This is at once the great methodological virtue of the cognitive model and its conceptual limitation. It is a virtue because it simplifies matters by off-loading subjectivity into the (nonscientific) domain of consciousness, ensuring that information theory can be safely applied. Subjectivity has been lumped under the heading of consciousness in order to get on with the business of modelling unconscious cognitive processes. And this strategy has proven highly effective over the past 50 years and more. Seen in this light, the hard problem of consciousness and the explanatory gap are symptoms of the cognitive model. It is an inconvenient truth, however, that some unconscious mentation *is* subjective. Subjectivity leaks into the unconscious arena.

Another historically prominent approach to unconscious thought may be characterized as "top-down" rather than bottom-up. The *psychoanalytic model* defines unconscious thoughts with respect to conscious ones. It uses psychological concepts, rather than information science concepts, to define the content of unconscious mental states. In the psychoanalytic model, the term "representation" has a slightly different meaning than it does in cognitive science . For most cognitive scientists, a representation is a syntactically defined data structure in an information system. Neuroscientists use "representation" in a similar way to mean a trace in the brain. In these uses of the term, a representation is a subpersonal (or third-person) element that need not be available to consciousness, even in principle. For the psychoanalytic and phenomenological traditions, however, a representation is a first-person thought that can in principle become conscious. The retinal image is a representation only in the former, cognitive sense. It is an image *in* me but not an image *for* me. I cannot see my retinal image, and no amount of meditation will allow me to. Contemporary cognitive theorists tend to think of unconscious emotions in the same way. An unconscious emotion is conceived as a third-person causal process in the head that helps to explain behavior, but it is not something that bears any subjective significance. The psychoanalytic model is top-down in that it begins with the first-person perspective *for* which the emotion is implicit, and then digs for the subjective salience. I use "representation" primarily in its cognitive sense, and mark the psychoanalytic sense with the qualification "subjective" or "for-me."

Psychoanalytic theory is often characterized as *depth psychology*. Freud (1925) captured this idea by way of analogy with an old toy – a writing pad in which the top layer may be wiped clean but the traces remain in the layers beneath. The psychoanalytic unconscious is the domain of embodied personal meaning that, like prototypically conscious experience, is part and parcel of subjectivity. For the purposes of this essay, Freud's most important thesis was that *subjective* thought is often unconscious.[23] Although the particulars of Freud's metapsychology are obviously wrong in a variety of ways, a concept *like* that of the psychoanalytic unconscious – call it the *first-person unconscious* – is still important for philosophy of mind. The first-person unconscious is structured by *baseline subjectivity*, the animal's affective engagement with the environment of its concern. Baseline subjectivity is the topic of Chapters 6 and 7 in this essay.

To clarify: the first-person unconscious is not Freud's unconscious. Like the traditional Lockean definition of consciousness, received conceptions of the unconscious (both cognitive and psychoanalytic) need to be amended and updated. For Freudians, unconscious drives have nothing to do with the brain. Rather, libido is the language of an *inner self*, a psychosexual language that, although animated by dynamic drive, is always a matter of symbolic self-expression. On that model, "sexuality is the hermeneutic adventure of psychic energy" (Marabou, 2012, p. 36). Here, the phrase "psychic energy" is used in strict contrast with *electrochemical* energy. For a traditional Freudian, psychic energy is itself a representation, and it exists only within the mental world of a true inner self. This psychosexual symbolism is what gets "repressed," diverted, rerouted, and reconfigured in the unconscious.

There is no true inner self. The first-person unconscious is not a *self* at all, and it is neither truer nor more inner than the experiences of which we are consciously aware. And in its baseline form, it is not linguistically (symbolically) structured. The first-person perspective is a coordinated affective relation with the environment, and the first-person unconscious is just the subset of these for-me states of which we are unaware. Catherine Marabou speaks of a *cerebral unconscious*, enacted by affective self-regulation (Marabou, 2012, p. 39). She helpfully characterizes this cerebral auto-affection as *nonreflective*.[24] The affective brain represents information in a way that is self-referential, but is not a self-representation. In the homeostatic regulation of affect, Marabou suggests, the embodied brain can feel itself as it lives through salient changes in its environmental situation, using these affects as governors. This cerebral unconscious "represents itself without presenting itself...*Cerebral auto-affection is the unconscious of subjectivity.*" (pp. 44, 43). In Chapters 6 and 7 I argue that this form of representation is a neural instantiation of identification free self-reference, the basis of the first-person perspective.

4.2 Unconscious subjectivity and emotion: toward affective neuroscience

Since emotions always matter, and mattering is always mattering for some subject, unconscious emotions are better understood in terms of the general psychoanalytic model than under the cognitive rubric. Unconscious emotions, if they are to count as emotions at all, must be subjective in much the same way that conscious emotions are subjective. At this point an example is in order. I present a typical case of the kind offered up by Freudians, ripe for a standard psychoanalytic interpretation in terms of repression. But the first-person unconscious it displays, I suggest, is more basic and so to speak more *ecological* than the standard Freudian depiction of the unconscious.

I relate a story about a friend from graduate school. I noticed an odd pattern in his behavior. Every day as he walked to campus, he took a circuitous route that added several minutes to the trek. One day as we walked together, I pointed out this odd behavior, which he had not previously noticed. He soon realized that he had been unconsciously avoiding his old neighborhood. He now noticed that he had negative feelings about that area. When I asked him why, he said he didn't know, but that he simply preferred the long way round and that he didn't like to walk past the old house. Later, as we walked on further, he recounted an episode that he considered to be ancient history. A few years before, his house in that neighborhood had been burglarized. We were puzzled – both about why he had not thought of it immediately, and also why that event should affect his peripatetic habits the way it evidently did. The next time we met, he immediately told me that he had been giving quite a bit of thought to the burglary and the unconscious avoidance behavior. Among the things stolen that day was a custom-built electric guitar, an heirloom he had recently inherited from his late older brother. It was one of the few things left of him, and utterly irreplaceable. Now the loss of the guitar had become a symbol for the "loss" of a brother. Furthermore, my friend explained that he had never wanted to think about his brother's death, and felt bad about it in a variety of ways. Now years later, he realized that he was avoiding the symbolic site of loss without any occurring awareness of doing so.

Consider the example as follows. Although mechanisms of the cognitive unconscious certainly entered into the causal explanation of the behavior, by themselves they did not shed light on its subjective significance. Rather, a person's way of life implicitly "refers back" to the individual. The emotion was potentially available to him all along, but was not explicitly known. Instead it was something that he lived through, implicitly shaping his perceptual world. His habitat or turf (*terroir*) acquired an affective valence for him, such that his subjective space was invested with value. The person's orientation to the environment was governed by a level of emotional engagement that was simply there, unacknowledged. And although the case was later narrated by the reflective subject, the actual first-person way of life was simply caused by a sequence of events, not by a narrative rehearsal of them (see Chapter 2 for more on narrative). Emotion then, is ecological at base. My friend's world came to be infused with idiosyncratic first-person content, a belief-desire structure, linguistic and symbolic mediation, and practical import for his personal life. But no specifically Freudian baggage is involved in the first-person unconscious – no theory of repression, no neurosis of personality, no covert agency or hidden conflict between ego, superego, and id.

The neuroscience of emotion is contested ground. So-called *Affective Neuroscience*, championed by Jaak Panksepp (1998, 2005) and embraced in this essay (see Chapter 6 for a more complete assessment of affective neuroscience), takes a thoroughly subjective and first-person view of limbic function, grounding subjective emotions in the action of dynamic systems in and across midbrain and basal forebrain structures like the amygdala and the periaqueductal grey. Affective neuroscience has much to say about how subjective engagement shapes the animal's lived world. But Panksepp's view of affect is still the minority report. Several more mainstream neuroscientists emphasize nonsubjective unconscious processing in pursuit of a cognitive theory of "the self" (LeDoux, 1998 & 2002; Damasio, 1999 & 2003). Not surprisingly, Joseph

LeDoux approaches both emotion and the self from within the cognitive model of the unconscious. He writes:

> Cognitive science was successful because it figured out how to study the mind without getting bogged down in questions about subjective experience. The trick was to treat the mind as an information processing device rather than as a place where experiences occur... [And the] processing approach is, in fact, directly applicable to the study of emotion. (LeDoux 2002, p. 205)

LeDoux elaborates that aspects of emotion that have been taken as constitutive by psychologists and philosophers alike, such as subjective feeling or intending, are inessential. His goal is to use an information processing conception of emotion as a basic element in a reductive model of the self. Thus, he characterizes the self as a synaptic structure: "You are your synapses. They are who you are" (LeDoux, 2002, p. 324).[25] His account of the self and emotions, then, fits perfectly with the cognitive model of unconscious thought.

It is true, in a way, that your synapses are *what* you are. But consider that the point of psychoanalysis, like that of reflection more generally, is the development of one's character. It asks the question: "What *sort* of person am I?" (Rorty, 1991, pp. 152–3). This is an act of qualitative self-interpretation that involves becoming aware of on one's subjective world. The answer to the question "What sort of person am I?" is not discoverable from facts about the brain alone.

To see this, consider a prominent philosophy of emotion that tries to combine the cognitive model of unconscious thought with the sort of "first-personalism" about emotions I have in mind. Martha Nussbaum (2001) adopts a neostoic approach on which emotions are understood as judgments about personal flourishing (*eudaimonia*). Her view has many strengths, one of which is this: Nussbaum holds that although the existence of unconscious emotions indicates that there is no necessary phenomenological condition for emotion, nevertheless unconscious emotions still retain their subjective status. Accordingly, Nussbaum is at times sympathetic to some form of depth psychology, but one that is free of problematic Freudian baggage such as the theory of repression. (Nussbaum, 2001, pp. 61, 64, 147, 71, 181).

Nussbaum assumes that the cognitive model remains allied with her eudaimonistic philosophy of emotions, and this enables her to cite LeDoux, among others, in support of her view. Nussbaum, 2001, p. 114). But her account of personal flourishing conflicts with LeDoux's stated method of not getting "bogged down" in questions about subjectivity. There is a slippage between the subpersonal processes adverted to in the cognitive model of the unconscious and the critical notion of eudaimonia central to Nussbaum's account. Insofar as eudaimonia is the reflective project of an embodied subject, it is indeed essential to emotion. Personal flourishing is not just the relative fitness of a neural mechanism in a past reproductive environment. Subjectivity is required for eudaimonia, just as it is for conscious awareness more generally.

5 Conclusion

The assumption that all subjectivity is conscious is unwarranted, and philosophical psychology requires a conceptual space for unconscious subjectivity. Progress can be made by distinguishing subjectivity from conscious awareness, and recognizing that the former is necessary but not sufficient for the latter. This will render unconscious subjectivity useful for philosophy of emotion, and at the same time preserve an empirically tractable conception of consciousness as awareness.

4
What Subjectivity Is Not

1 The paradoxical duality of qualitative awareness

1.1 Introduction

An influential thesis in contemporary philosophy of mind is that subjectivity is inexplicable because it is paradoxical. Joseph Levine's *Purple Haze: The Puzzle of Consciousness* (2001) made this thesis explicit and connected it with the so-called "hard problem" of consciousness (Chalmers, 1996). The peculiar claim that there is a paradox about subjectivity arises within a certain tradition and a particular set of texts, and the ways they encourage philosophers to think about the topic. In this literature, subjectivity is conceived as *inner awareness of qualia*. Drawing on this conception, Levine (2007, 2001) has argued that this unique form of awareness generates a paradox resisting empirical explanation. Levine concludes that the hardest part of the hard problem of consciousness is to explain *how anything like a subjective point of view could arise in the world* (2001, p.174). Against this, I argue that the nature of subjective thought is not correctly characterized in terms of inner awareness, and that the problem about the subjective (first-person) point of view should be distinguished from the perennial metaphysical problem of qualia or phenomenal properties.

Subjective consciousness is always "first-person" in that it is *for* some subject (Kriegel 2009; Zahavi, 2005; Levine 2001; Shoemaker, 1997; Nagel, 1974).[1] In Chapter 2 I argued that the for-me structure of the first-person can be understood as a variety of *identification free self-reference* (Evans, 1982). My aim in this chapter is to open space for such an account by arguing against the paradoxical but widely held alternative, in which subjectivity is construed as a privileged and intimate relation to qualia.

It is somewhat unusual for contemporary philosophers to countenance such a thing as a paradox. Philosophers today are more likely to consider sets of incompatible claims, or to speak of an explanatory gap in our understanding. But framing subjectivity as paradoxical is really only an acute way of pointing to a conceptual problem. Speaking of a paradox helps convey a sense of wonder about something beyond our ken, something that seems impossible but true. A paradox is a particular kind of mystery in which incompatible truths collide. One way to say this is that a paradox arises when there is a sound argument for a manifestly false conclusion; for example, the conclusion that the hare can never overtake the tortoise. In such a case it appears that both P and ~P are true. This is an intolerable situation that is not just unexplained but really *inexplicable*. There can be no progress toward an explanation because no putative framework could possibly explain both P and ~P, on pain of inconsistency. In the teeth of a paradox there are four basic responses. First, one can "explain" it by adopting a special ad hoc ontology (in this case, dualism). Or, stoically, one can accept the inexplicable nature of things (as in mysterianism). A third kind of move is to dissolve the paradox by changing the way we talk about P, and the way we worry over paradox, such that it no longer seems so upsetting (philosophical behaviorism, neo-pragmatism). Finally, one might try to *resolve* the paradox. This means rejecting either P or ~P, more or less directly. With respect to the nature of our first-person relation with our own experience, this paper makes the fourth response.[2] There is no paradox about subjectivity. This is not because there is no *problem* of subjectivity, as the third "therapeutic" strategy would have it. There is both a problem and the form of a solution. But this has not been the prevailing view in the literature.

Section 1 rehearses the paradox, discussing it in conceptual, phenomenological, and historical context. Section 2 suggests how to handle the first-person perspective in a non-paradoxical way, comparing subjective judgments about qualia with similar judgments about episodic memories. Section 3 considers and rejects a relevant alternative interpretation of our subjective relation to the contents of experience, known as the Transparency Thesis. Finally, Section 4 considers a set of objections based in the Phenomenal Concepts Strategy.

1.2 The paradox

To clarify how subjectivity is meant to be paradoxical, it is simplest to begin by briefly introducing qualia and the explanatory problem that attends them. The term "qualia" refers to the *feels* of experience – the redness of red, or the sweetness of honey (I will sometimes use the phrase "phenomenal property" in an equivalent way). Today's materialist science of the mind understands thoughts as representations, and representations as information functions.[3] In this context a persistent problem about qualia arises because, although encoding information about the world can be given a functional (and hence a material) analysis, this seems to leave out the qualitative aspect of the experience. Consider, for example, that the words "the tomato is red" carry the information about the color perfectly well but do not carry the qualitative content. The sentence itself is not red. Similarly, mechanical sensors can be built (or nervous systems can evolve) to detect the relevant part of the spectrum but not the *redness*. Qualia per se don't seem to serve an information function, and there seems no adequate paraphrase for qualia in purely physical (functional or causal) terms. This is vexing for materialist philosophers of mind, and qualophiles such as Levine and Chalmers are surely right in arguing that we will not have a physicalist explanation of consciousness as long as this problem persists.

But this problem about qualia is not the paradox of subjectivity. It is rather a problem of *objectivity*; that of explaining how there can be such things (properties) as qualia in a physical world. The paradox concerns the first-person mode of experience in which we encounter these qualia. In effect, philosophers of consciousness have appealed to Russell's (1911, 1914) notion of knowledge by acquaintance to stake a claim that qualia are *directly grasped*. If you "zero in" on the qualitative character of your experience, it is said, you don't seem to register the same cognitive relation that you find when focusing on other, non-experiential or merely conceptual forms of mental content.[4] If you think of your coffee cup, for example, as "that thing you carry with you to lecture and often use as an example," this is a conceptual <u>description</u> that can exhaustively specify your cognitive relation to it and the role it plays in your mental life. By contrast, it is urged that your relationship with qualia – the color of the cup, say – is more intimate than this. Qualia are held to have a certain *presence* in experience that mere conceptual descriptions do not. Levine writes:

> ...it seems the right way to look at it is that the reddishness itself is serving as its own mode of presentation...the reddishness itself is somehow included in the thought; it's present to me. (2001, p.8)

Levine calls this the *substantive* aspect of subjectivity, which is one of the principle ways in which one's subjective grasp of what-it-is-like eludes explanation in cognitive or semantic terms (*ibid.* p.9). In addition to being substantive, the intuition of qualia is also said to be *determinate*. This means that subjective grasp is sufficient to render a qualitative character like "reddishness" identifiable independently of its relation to other qualities. That is, there is something <u>intrinsically</u> reddish with which the subject is directly acquainted, such that redness is immediately intuited. By contrast, the way you grasp the property of being a coffee cup consists not in your intuition of the essence of "coffeecupness," but rather in representing various relational features of the object before you, such as being heat resistant, hand-sized, and with an opening for drinking.

The paradox of subjectivity concerns this claim that the subjective mode of presentation consists in a direct relation with qualia. Levine argues that his conception of subjectivity as the direct, substantive, and determinate grasp of qualia flows from the nature of qualia themselves. The fact that experience is qualitative, it is argued, pushes us to conceive subjectivity as the *necessary awareness* of this qualitative nature. The attempt to distinguish these elements from one another, to analyze the problem of consciousness into constituent problems, will only lose the phenomenon of interest. Thus, according to Levine there is a "paradoxical duality" of experience, in which qualia and our awareness of them are the inextricably entwined relata in an unanalyzable relation:

> There is an awareness relation, which ought to entail that there are two states serving as the relata, yet experience doesn't seem to admit of this sort of bifurcation...Qualia are such as to necessitate awareness of them, and certain thoughts seem to include qualia in their modes of presentation in a cognitively special way. (Levine, 2001, p.168)

Thus Levine identifies subjectivity with a special *awareness of qualia*. The paradox is that qualia are <u>both</u> the objects of knowledge and also part of the knowing itself. This arises because there is a *presentation relation* between qualia and the subject but this relation is also an indissoluble unity, internal and intrinsic to subjectivity, and hence cannot be a relation between two separate things. So, subjectivity presents us with a paradoxical duality. This view of subjectivity represents an attempted generalization of the hard problem of consciousness, from a problem about the nature of qualia to a further problem about *inner awareness* of qualia.

How does this situation present a paradox? That is, how can it be reconstructed as a sound argument for an apparently false conclusion? The reasoning is roughly as follows (and I think Levine would accept this formulation). According to prevailing materialist philosophy of mind, awareness has a cognitive dimension; to be aware of your experience is to represent it in some way. Representation is always externally related to that which it represents, in the sense that the representation can malfunction and misrepresent, or can be destroyed, independently of what is represented. Another way to say this is that for any representation R, R is correct iff x, where x is some independent fact about the world (cf. Hellie, 2007). So materialist philosophers of mind have sound reasons for the conclusion (P), that awareness and qualia are distinguishable aspects of a relation between representation and what is represented. So far, R is the awareness and *x* is the quale, an independent fact. But this is paradoxical, it is argued, because we know this conclusion to be false. This latter claim, (~P), is grounded in the doctrine that qualia essentially include our awareness of them. That is, in virtue of their "presence" in consciousness, qualia *necessitate* awareness in a way that mundane objects of representation do not. Since qualia are (said to be) intrinsically present in the special way, this requires that they cannot exist apart from our awareness (representation) of them. So, P is false; qualia and awareness are *not* distinguishable aspects of a representation relation. Subjective awareness, then, is an inner and intrinsic knowledge that seems to require a special ontology. Since materialists do not wish to admit a special ontology, they are left with a paradox.

In writing about this paradox of subjectivity, Levine relies as much as possible on ontological-sounding locutions, for example, about our "grasp" of qualia, about the "intimacy" of our relation with qualia, and about the "presence" of qualia to consciousness. He explains in a footnote that he avoids Russell's vocabulary of knowledge by acquaintance because he doesn't want his work to be confounded by the problems associated with Russell's epistemological project (Levine, 2001, p.179, fn.8). But the epistemic problem is not easily deflected.[5] An anonymous referee (like many parties to this debate) denies that the paradox of subjective duality carries any epistemic implication. All that is required, it is often said, is that we be "aware" of qualia, not that this awareness constitutes knowledge. But this response renders the paradox vacuous. If subjective inner awareness doesn't require us to be aware *of* any particular content, then it is an empty designation. All of the talk about substantive, direct, and determinate inner presence is rendered toothless. Our "awareness" of qualia then amounts to nothing at all; asserting it is just another way of saying *that* there are qualia. Or again, if there is no cognitive component to "awareness" then (P) is false and there is no reason to conclude that there is a paradox of subjective duality. But if inner awareness retains its paradoxical bite, then it is false that it carries no epistemic implication. For recall that the claim of the inner awareness approach is that subjective thoughts "include qualia in their mode of presentation in a *cognitively* special way" (ibid., emphasis added).

The epistemological significance of the identification of subjectivity with the awareness of experience has been repeatedly emphasized by philosophers since at least Descartes (2008 [1641]).[6] The inner awareness conception entails that subjective judgments about the qualities of experience are infallible in a way that no other synthetic judgment can be. If it is true that subjectivity is yoked to qualia in these direct, substantive and determinate ways – if qualia are actually present to us – then there is no mistaking their intrinsic nature. The term "intrinsic" here is important. It is what is held to be special and inexplicable. If awareness is <u>not</u> intrinsic to qualia (and if it is not awareness *of* their intrinsically qualitative content), then their cognitive mode of presentation is not distinctive in the ways requested.[7] Thus, when a subject carefully attends to the phenomenal character of experience and judges it to be "reddish," this judgment cannot be made in error because the redness partly constitutes the judgment. The judgment is not merely constituted by the feeling that there is some quale or other. It is <u>determined</u> by the red content. Although Levine might try to disown this implication, it is it near the heart of his view about subjectivity. For it is in virtue of this special epistemic privilege that subjectivity is supposed to present an *additional* problem for materialism, beyond the ontological problem about qualia. And Levine is very clear that he considers subjectivity to present this additional paradox (Levine, 2001, pp.167–174). The new problem arises at the epistemic level because of its incompatibility with a fundamental commitment of the representational theory of mind: namely, that our representations can always be mistaken.

Now the problem is in full view. The purportedly unique kind of inner awareness that gives rise to the paradox is rooted in traditional philosophical ideas about the nature and status of subjective experience. To give it a name, we might call it the *Russellian Acquaintance* model of subjectivity.[8] The acquaintance model is fundamentally grounded in a distinction between sense data (qualia with which we are directly acquainted), and conceptual description of everything else. Only on the basis of Russellian acquaintance can it be concluded that no purely cognitive approach (and hence no materialist approach) can account for subjective inner awareness. Thus, if Russellian Acquaintance is the wrong model, there will be no basis for holding that subjectivity is a special paradox as inexplicable as qualia themselves. The problem about qualia will remain, but the first-person perspective will cease to be so mysterious. The first-person perspective can then be treated as a distinctive but not paradoxical form of mental representation.

1.3 Inner awareness in historical context

As we have just seen, the thesis that subjectivity is paradoxical rests on the idea that the mode of presentation of qualia to the inner eye is direct, substantive, and determinate such that one is necessarily aware of an intrinsic content. Where did this idea come from? The inner awareness doctrine is usually presented as a discovery of phenomenology or introspection, an argument which I consider in detail in Section Two, below. But it also has a history. It arises from particular ways philosophers have thought about subjectivity throughout the modern period and the twentieth century. In reflecting on the contemporary debate, a brief detour through (some of) the historical background will be useful in reframing the issues for the rest of the paper.

The paradox of subjectivity is explicitly framed as a challenge for materialism, so it may be helpful to consider it in light of idealist metaphysics and epistemology. Begin with Berkeley's venerable thesis that *esse* is *percipi*. Of course inner awareness does not require that, *in general*, to be is to be perceived. Levine's claim is restricted to one particular realm of being, the realm of qualitative presence. That is, proponents of inner awareness hold that qualia are those entities whose being is essentially bound up with perception, and subjectivity is that mode of presentation whose being is perceiving. Bishop Berkeley's slogan is wonderfully simple, endlessly thought-provoking, and notoriously difficult to refute. I certainly won't refute it here.[9] But reflection on this slogan will help clarify what is going on in the paradox of subjective duality, and indicate the way toward a non-paradoxical account.

How should the slogan, *esse* is *percipi*, be understood? I adopt the following interpretation, found in Moore (1903, p.438): What makes a thing real is its presence as an inseparable aspect of sentient experience.[10] For the purpose of this discussion, the next step is to restrict the range of this formulation to the domain of qualia, yielding: *What makes qualia real is their presence as an inseparable aspect of experience.* Here is the core idea that gives Levine's paradox its legs. Moore's language here is a bit old-fashioned, and contemporary philosophers are unlikely to claim explicitly that presence to awareness is what "makes qualia real," but this is entailed.[11] Recall that according to the paradox of subjective duality, qualia necessitate our awareness of them. So awareness is a necessary condition for qualia. Notice that although this is not a causal claim neither is it just a claim about conceptual priority. It is not merely the claim that that the *theoretical concept* of qualia is derived from the concept of awareness, such that one cannot have the former theoretical concept without having the latter. This would be the manner in which, for example, mid-sized objects are conceptually but not ontologically prior to subatomic particles (Parfit, 1999). It is true that one cannot conceive of a proton without the prior concept of an object. But this does not generate a paradox for physics. Similarly, even if it were true that the concept of qualia depends on the concept of awareness, this would not generate any paradox for materialism. To present a paradox for materialism, awareness must be taken as an ontological condition for qualia: in order that there be qualia, it is necessary that there be perception (awareness) of them. The upshot would then be that subjectivity must be understood as the necessary "perception" of qualia.

According to the above, to be perceived is to be "present as an inseparable aspect" of experience. Modern empiricists from Locke and Berkeley to Russell took it that what are immediately "present" in perception – that with which we are immediately acquainted – are *sensations*. Few philosophers or psychologists today would accept that perception is composed of sensations, sense data, or any other basic unit of consciousness. Nevertheless, "qualia" in the contemporary literature are the descendants of modern philosophy's "sensations," i.e., qualitative conscious elements (cf. Smart, 1959). It is these that are said to be present in the inner mode of awareness. The old theory that perception consists in sensations sustains an obscure understanding of "awareness" as both an epistemic relation and as an inner aspect of the perceiver.

G.E. Moore (1903) attempted a clarification. He argued that a qualitative content such as "blue" can be related to consciousness in two ways. Blue content could be either an *object* of consciousness or it could be a *property* of consciousness.[12] It will be readily seen that this distinction closely tracks the two sides of Levine's paradox. A blue sensation considered as an object of consciousness is something that is known, an intentional object that the subject might or might not have been aware *of*. But sensation considered in the second way is very different. Here the blue content is an *aspect* of experience, in the same way that roundness is an aspect of a circle, "present" in the circle. This tracks well with the contemporary idea, favored by Levine, that blueness is *part* of my experience. Thus, *awareness of blue is blue*. As Moore observed, the consequence is that a blue sensation is <u>not</u> a sensation of or about blue! The relation between subjects and their qualia is no longer an intentional relation at all. This is an unfortunate result, for it is undeniable that in judging "I see blue," I form a contingent knowledge claim in which "blue" figures as the object of an intentional state (i.e., as the content of a representation). I do not merely *express* my necessary blueness, for such an expression would be neither true nor false and would not constitute a claim at all. This shows that inner awareness ("percipi" in Berkeley's language) must be understood to involve qualia both as objects and as properties. So another way of expressing Levine's paradox is to say that subjectivity requires qualia to play both roles *at the same time*.[13] To anticipate, I hold that qualia are <u>first</u> properties of the subject and only <u>later</u> objects of awareness, and that the existence of the former is not sufficient for the occurrence of the latter.

Esse is *percipi* is part of the historical background for the paradox of subjective duality, supporting the ambiguous use of "immediacy" as having both a phenomenological and an epistemic sense. This ambiguity is damaging because on one of the two meanings but not the other, it is *true* that first-person experience is immediate. Qualia are always someone's qualia, and it is true that for sensations to exist they must be sensed somehow or other. So at this stage the *esse* of qualia really is *percipi*, where this is construed to mean that qualia are the properties of some subject. Subjects must directly register, be affected by, or inwardly express their qualia. In this sense, qualia are *phenomenologically immediate*. There need be no further representation, interpretation, or reflection in order for qualia to <u>be</u>, or in order for them to *be subjective*. From here, however, it does not follow that subjects are uniquely and intimately aware of their qualia, except in the empty sense that, e.g., a blue flower must also be "aware" of its blueness, a circle must be aware of its roundness, or a Leibnizian monad must always "perceive" the entire universe of relations to it.

So much for the historical background. From it I have extracted two distinctions relevant to the contemporary debate, one between phenomenological and epistemic immediacy and another between the properties and the objects of experience. I now return to the paradox of subjective duality.

2 Inner awareness beyond the paradox

The view at issue is that qualia, by nature, constitute an inner awareness *of themselves*; i.e., an awareness of an intrinsic subjective content.

Why accept this? The reason is intuitive, presented in good phenomenological form: *it is there*. Direct knowledge of qualia is supposed to be a simple fact of experience, and the distinction between this and descriptive knowledge is verified through introspection (or revealed through phenomenological reflection on the first-person mode). Call this the "immediacy intuition."[14] The immediacy intuition is hard to shake. But although this fundamental and intuitive reason for thinking of subjectivity as privileged inner awareness is psychologically compelling, it is not epistemologically conclusive. It is possible to accept that subjectivity is phenomenologically immediate (lived, embodied, or first-person), without also accepting that this fact grounds any knowledge by acquaintance, direct grasp, or inner awareness. Thus, while I agree that there is a first-order level of experience that is directly lived by the subject, I deny that the content of subjective experience is immediately available.[15]

What, then, is the nature of our first-person relation to our own qualia? In this section I suggest a simple approach in terms of *incorrigibility* rather than infallibility. The epistemic incorrigibility of judgments about subjective experience does not derive from the direct, determinate, or substantive qualities that are said to accompany introspective awareness. Rather, it results from the simple fact that individual experience is not accessible to others. The first-person perspective ensures that others cannot get "inside" your particular subjectivity. This section is offered by way of providing some glimpse of what subjectivity looks like when inner awareness is jettisoned. This is intended to undermine some of the prior motivation for adopting the inner awareness view. Broad reflection on what is desired from an account of subjectivity indicates that inner awareness is blinkered and inadequate to the phenomenology of the first-person.

Put aside for a moment the peculiar ontology of qualia. There are many features of first-person experience which do not fit the model of substantive and determinate presence. There are unfocused, liminal, unattended, peripheral, and background aspects to the stream of consciousness. There are poorly understood experiences. There are dreams. Most of subjective life is fleeting and vague, intangible and elusive. Of course, this conflicts with what I have dubbed the "immediacy intuition," to the effect that I cannot *help* having subjective knowledge because when it comes to subjectivity, the truth is self-revealing. But the domain of subjectivity – the purview of the first-person perspective – is wider, deeper and more elusive than might be supposed. Subjective mental life concerns much more than the momentary sensations that qualophiles fix on. Subjectivity also concerns happiness and other emotions connected with our personal well-being. There is a vast literature that takes this richer approach to subjectivity, and there is a complete disconnect between it and the literature on qualia. In this other literature there is widespread agreement that we are ignorant of much of our own subjective mental life and even ignorant of the character of occurrent affective states (Haybron, 2010; Nussbaum, 2003). If one takes seriously the idea that there are unconscious emotions, this means that subjective thought can even take place unconsciously (Neisser, 2006). That is, there are "for-me" emotional episodes that are very much like consciously experienced emotions, except that the subject remains unaware of them.

Perhaps all this merely changes the topic? One might argue that Levine and company are talking only about a very specific form of introspective awareness, and so should not be held accountable to other concerns about "subjectivity." But this response is self-sealing. Emotional experiences and dreams are first-person subjective mental states, and so they are fair counterexamples. If the inner awareness view makes a complete hash of them, it is rendered implausible on its face. Competing analyses cannot be required to concern themselves *only* with this one explanandum – stipulated as paradoxical – to the exclusion of other aspects of subjectivity. The point is that there is readily available independent evidence that for-me subjectivity is not best understood as inner awareness. Perhaps the first-person perspective does not essentially consist in a kind of *awareness* at all.

All this suggests a straightforward way of thinking about the epistemic status of introspective judgments about qualia. Introspection is a "private" or "dedicated" source of knowledge about my occurring subjective states. Since beliefs about contingently occurring states of affairs can always be mistaken, I can be wrong about my current qualitative subjective experience.[16] Thus, a deflated theory of privileged knowledge by acquaintance can be sustained by turning it into a weaker theory regarding a class of events about which individual judgments are incorrigible: *My acquaintance with my own experience cannot be corrected by others*. In 1970, Richard Rorty argued just this:

> Mental events are unlike any other events in that certain knowledge claims about them cannot be overridden. We have no criteria for setting aside as mistaken first-person contemporaneous reports of thoughts and sensations, whereas we do have criteria for setting aside all reports about anything else. (Rorty, 1970, p.413)[17]

Calling a judgment incorrigible is not at all the same as saying it is infallible. Incorrigible claims can be mistaken, they just can't be corrected. A nice consequence of this idea is that subjective incorrigibility becomes a relative matter. A given subjective report will only be more *or* less incorrigible, depending on the standards and techniques available to others. For example, since there are behavioral implications of holding beliefs and desires, our judgments about our own intentional states do indeed admit of correction by others (Rorty, 1970, p.420). Our access to our own subjective states comes in a continuum of incorrigibility, with judgments about occurrent and attended sensation events (i.e., qualia) at the maximally incorrigible end of the scale.[18] Judgments about liminal or background sensations are one step more corrigible, about emotions and beliefs another step down the line. Judgments about happiness or well-being are still more difficult to justify through introspection alone. Towards the far end of the spectrum lies episodic memory. By today's forensic standards, memory judgments are only minimally incorrigible. Eyewitness testimony just ain't what it used to be. If I am right in distinguishing subjectivity from introspective awareness, then there is a sense in which really *all* awareness is a form of memory. Consciousness is remembered subjectivity. This result is broadly in accord with the idea of "working memory" in the literature (e.g. Baars, 2002).

According to the present line of reasoning, the incorrigibility of subjective awareness is a matter of its *in*accessibility to others, not any special ontology of inwardness. Further, the substantive and determinate intuitions that accompany introspective awareness of qualia, much ballyhooed in the literature, are *just more phenomenal properties*, accessed reflectively. In principle the case is no different from others in which felt or intuitive certainty is no guarantee. Consider the following. It is now known that the felt certainty of a memory – its force, vivacity, and clarity – is not strongly correlated with its accuracy (Sharot, Delgado & Phelps, 2004). Even "flashbulb" memories of important or extraordinary events are often inaccurate, not just in detail but wholesale. Your memories of where you were when you first heard of the events of 9/11/2001 (say) may well be inaccurate in a variety of ways. Cherished memories of childhood events are almost certainly inaccurate. Even the memory of *how you felt* is not infallible. In a large longitudinal study, Hirst & Phelps, et. al. (2009) found that not only the content of event memory (for details of the event itself) but also the "flashbulb" aspects of what-it-was-like-for-me to experience the event are mediated by rehearsal and other community practices, not by intrinsic features of the first-person perspective. In short, it is a relatively recent discovery of psychology that first-person, episodic memory is not nearly as reliable as it seems.

But of course the first-person presentational mode often *feels* compelling! The intuitive conviction that a memory must be correct, for example, is based on the way it is present to consciousness; it is based on the "intimate grasp" of the memory as it is experienced. And this is exactly the same sort of evidence that defenders of acquaintance appeal to as the guarantor of their knowledge of qualia: "I know what happened! I can see it so vividly!"; "I'll never forget how I felt that day!" Under certain conditions, the subject may find it difficult or impossible *to doubt* the delivery of introspection. But this psychological compulsion, by itself, does not constitute knowledge. This example may help to defuse the "immediacy intuition," that the substantive and determinate form of qualitative experience is a guarantor of its content.

The basic point is that our unique first-person perspective on our own subjectivity is *usually* an overriding justification, rendering the judgment about the content of experience more or less incorrigible. But more generally, in order for the "determinate" feel of introspection to justify a judgment, it must be taken up and represented as a justifying reason (McDowell, 1994; Sellars, 1997, [1956]). All such justifications are defeasible and must "defer" to external and intersubjective standards whenever relevant and available. In Section 4 I pursue the implication of this last thought for the theory of phenomenal concepts. But first I must pause to consider the Transparency Thesis, which also descends from Moore (1903) and which purports to offer an alternative to the paradoxical analysis of our first-person relation to our own experience.

3 The Transparency Thesis

The previous section argued that subjectivity need not be considered as the paradoxical inner awareness of qualia. And indeed, the paradoxical analysis is certainly not the only contemporary construal of the relation between qualia and awareness. The so-called *Transparency Thesis* is a relevant position expressly denying that qualia are the properties of subjective awareness. Thus, it might be thought that this important and closely related thesis offers a constructive way to respond to the paradox. In this section I turn to consider this alternative, arguing that it is unappealing in its own right.

As noted in Section 1.3 above, the property/object distinction descends from Moore's (1903) "Refutation of Idealism." Today Moore's paper is much better known as the source of the so-called Transparency Thesis:

consciousness is "transparent" in the sense that its properties can never be introspected or perceived because consciousness is always intentional; it is always consciousness *of* something other than itself. According to the Transparency Thesis, qualia are known only *through* consciousness and are not "in" consciousness per se. Several writers have advocated the Transparency Thesis (a thesis they attribute to Moore) in an attempt to give a purely representational theory of consciousness (Tye, 2009, 2000; Dretske, 1995). The Transparency Thesis is opposed to the paradoxical inner awareness doctrine insofar as it holds that qualia are not "present" in consciousness at all but rather external properties out in the world (Tye, 2009, p.120). But the Transparency Thesis is subtly mistaken both as a thesis about consciousness and as an interpretation of Moore (cf. Hellie, 2007. See fn.18). Neither first-order subjectivity nor the second-order awareness of it is perfectly transparent. The upshot is that qualities are neither directly grasped in their intrinsic nature *nor* are they mind-independent properties represented in a transparent medium.

The Transparency Thesis gets its name from a passage in which Moore remarked that consciousness... "seems, if I may use a metaphor, to be transparent – we look through it and see nothing but the blue" (Moore, 1903, p.446). Moore is here treating consciousness as *that which makes the sensation of blue a mental fact*, and arguing that whatever this is, it is very difficult to point to or introspect. He observed that consciousness per se "seems to escape us."[19] The proponents of transparency interpret this observation strongly, holding that the properties of consciousness *do* escape introspection and that all we are ever aware of is an intentional object. After all, the representational theory of mind naturally suggests that to be "aware of" a thing is to be aware of the content of a representation. And to be aware of the content of a representation, it is argued, is precisely not to be aware of the representation itself (the vehicle). Phenomenal properties, then, are *what are represented* in consciousness. For example, the colors of the tapestry before me are experienced as properties of the tapestry, not as features of the experience (Tye, 2009, p.117). In Moore's terms (Section 1.3, above) the Transparency Thesis amounts to an identification of qualia as the objects of consciousness and a corresponding denial that they are properties of the subject. The blue sensation, then, is not itself blue.

The Transparency Thesis is appealing to physiologically-inclined philosophers because it exorcises qualia from the mind, thereby turning phenomenal properties into Somebody Else's Problem.[20] If phenomenal properties are mind-independent, then the inability to explain them is no longer a problem that must be faced by the representational theory of mind.[21] But this only makes things weirder; qualia are now said to exist out in the world, *without a subject* to have them! Thus, the Transparency Thesis only passes the buck down the line to physics. Michael Tye explicitly concludes that the Transparency Thesis renders the nature of phenomenal character "compatible with physicalism *provided that the external qualities themselves have a physical nature"* (2009, p.122, emphasis added). That is, qualia are Somebody Else's Problem. But phenomenal or qualitative blueness (e.g.) is no more a part of the lexicon of physics or optics than it is part of neuroscience or information theory (cf. Prinz, 2012). Contrary to the Transparency Thesis, it remains plausible that qualia are what-it-is-like for creatures like us to embody certain sensory processes. Here, Levine and the other qualophiles are on stronger footing, arguing that phenomenal properties belong to the first-person perspective, a mode of presentation that is not translatable into the language of physical facts (cf. Block, 2006). Sensory processes surely carry information about the world, and various kinds of representational content can be extracted from them. But in the first instance, phenomenal contents are not *themselves* the properties for which they serve as proxies. The mind-independent set of properties that causes phenomenal blueness is not *blue*. Phenomenal blueness is a stand-in for an external state of affairs, which it indicates.[22] If so, the Transparency Thesis is in trouble. And this makes sense, because of course representations are not in general transparent. In reading a text which says that the tapestry is blue, one does not look transparently through the representation to the tapestry (cf. Prinz, 2012, pp.14–15).

According to Tye's recent statement of the Transparency Thesis, I become aware of what-it-is-like to experience blue by being aware of the blueness of the tapestry before me (Tye, 2009, p.117). Unfortunately, this only smuggles the conflation of property and object back into the argument. This is because what-it-is-like to experience blue is not a property of the tapestry, but a property of the experience. And what-it-is-like to experience blue is also not the content of my experience *of the tapestry*. The content of that experience is that the tapestry is blue, and this is a content I can entertain even without knowing what-it-is-like to experience blue. Hence, I do not become aware of the property of the experience (what-it-is-like) merely by knowing something about the tapestry. In order to become aware of what-it-is-like, I must make my subjectivity itself the object of my awareness. It is for this reason that we have no Russellian acquaintance with sense data. The double duty of being both the property of a sensation and the representational object of awareness cannot be performed by one and the same thing at a single time. Ironically, this problem persists even for the Transparency Thesis. If experience is transparent and qualia are mind-independent, then these qualia certainly cannot deliver knowledge of what-it-is-like to experience them.

One final comment will conclude this section. Although the Transparency gambit fails, the paradoxical property/object identification need not simply recur at the level of introspection. I do not propose to rescue subjectivity from paradox only by postulating an "introspection module" that has the paradoxical inner awareness. That is roughly the problem faced by the advocates of the theory of phenomenal concepts (and the self-representational approach), discussed in Section 4 below. Instead, there just is no such special awareness. Once the doctrine of special inner awareness is relinquished (i.e., ~**P** in our paradox), the sound argument for the cognitive/interpretative nature of the first-person perspective (**P**) can be embraced, despite its counterintuitive conclusion that qualia and awareness can be dissociated. Thus, there is no attendant paradox of subjective duality.

4 Objection from phenomenal concepts

This chapter argues a negative thesis about subjectivity; i.e., that it is not the intrinsic inner awareness of qualia.[23] Even the positive discussion of incorrigibility in Section Two, giving an indication of what subjectivity might look like once inner awareness is abandoned, is included only as another way of making the case that the epistemic phenomenological immediacy does not deliver epistemic immediacy and that paradoxical view is unmotivated. But it may be objected that none of this is to the point because there are existing accounts, not yet considered, which demonstrate that the requested features of inner awareness can be had after all. Thus, the burden is on me to provide reason for thinking that my approach is preferable to those already on offer. In this section I argue that the relevant alternatives terminate in a dead end. I group this family of existing approaches together under the general banner of the *Phenomenal Concepts Strategy* (cf. Tye, 2009). As will be seen, I construe the Phenomenal Concepts Strategy in a way general enough to cover some theories that travel under different names. In particular, I treat the *Self-Representational* approach as a variant of this strategy because it shares the key idea that subjective awareness is constituted in a representation that takes itself as its own content (Kriegel, 2009, 2005).

Proponents of phenomenal concepts aim to provide an account, compatible with physicalism, of the special kind knowledge gained through experience.[24] The first move in this strategy is (once again) to distinguish between knowledge by phenomenal acquaintance and knowledge by standard descriptive concepts.[25] Descriptive concepts are *deferential*, governed by linguistic communities and experts. Individual applications of those public concepts can be mistaken and corrected by others. It is in virtue of this intersubjective nature that deferential concepts can count in knowledge. Qualitative categories are deferential; if I judge the wall to be painted blue, I can be wrong and I can be corrected. The wall is actually periwinkle, but I have not mastered this concept. Here, the error is based in my ignorance of a public classification scheme. But, it is argued, knowledge of qualia is not deferential in this way. Instead, my inner awareness of what-it-is-like is founded in another kind of concept altogether, a *phenomenal concept*. Roughly, a phenomenal concept is a demonstrative recognitional representation. It is demonstrative in that it takes the form *"that x,"* where x is a phenomenal quality. It is recognitional in that it connects *that x* to prior encounters with x. When I see the color of the wall, I see *that color,* and introspection is just the capacity to form judgments using phenomenal concepts.

On the phenomenal concepts strategy, the quale serves as its own representation. Rather than deferentially naming or describing, a phenomenal concept is supposed to pick out its referent by *exemplifying* the property and then pointing at it. In this way I can be mistaken about the public (deferential) classification but correct about the demonstrative indication.[26] Even when I am wrong about whether the wall is correctly categorized as blue or periwinkle, I am not wrong in knowing my experience as *that x*. My phenomenal concept of "that x" is constituted by a *sample* of x, so in order to acquire and employ the phenomenal concept, I must actually token the qualitative property to which it refers. Phenomenal concepts, then, are another theoretical apparatus meant to do double duty as both property and content of consciousness. My acquaintance with qualia is meant to ground a sort of demonstrative knowledge, constituted by what is known, referring to itself *as* what is known.

For the purposes of this discussion, Kriegel's (2009, 2005) Self-Representational theory of subjectivity can be treated as a closely related variant on the Phenomenal Concepts Strategy. According to Self-Representationalism, a mental state is subjective when it comes to constitute an intrinsic awareness of itself. In particular, Kriegel argues that inner awareness is a matter of the demonstrative self-indication of a real, instantiated phenomenal property (2006, pp.42–43). The key point shared with the phenomenal concepts strategy is that the property must represent itself *as itself* in the particular phenomenal content that it instantiates; i.e., the representation and the thing represented (the content) are supposed to be one and the same. My arguments against the phenomenal concepts strategy and, more importantly, against the paradox of subjective duality, also apply to Self-Representationalism, mutatis mutandis.

The significant difference between phenomenal concepts and Levine's paradox of subjective duality is that the phenomenal concept is often thought to be generated through the act of introspection, not by the quale itself.[27] On the phenomenal concepts strategy, then, subjective qualities can remain unknown as long as they are not introspected. So this is not quite a version of Berkeley's *esse* is *percipi* because introspective awareness is not a necessary condition for the very existence of qualia. This is an important step in the right direction. Nevertheless, when you acquire a self-indicating phenomenal concept through the act of introspection, it cannot fail to be a sample of itself. Thus, phenomenal concepts (not phenomenal properties by themselves) remain a sufficient condition for the "presence" or intimate grasp of qualia, and the paradox returns at the introspective level. Finally, if it is possible to construct a version of the account on which the phenomenal concept *can* misfire or fail to be properly sorted, the strategy loses its traction as an objection to my reading of first-person knowledge in terms of incorrigibility, offered in Section 2 above.

I argued earlier that qualia cannot do the sort of simultaneous double-duty that inner awareness requires, qua object and qua property of subjective experience. Now we are in a better position to see why. Recall the Sellarsian idea that in order for an experience to justify a knowledge claim, it must be taken up and conceptualized *as* the justification. It is precisely for this reason that phenomenal *concepts* (over and above phenomenal properties themselves) are introduced in the account of introspective knowledge. They are supposed to be what does the work of sorting experience into transportable and generalizable categories for cognitive use. But no token experience or quality can function as a stand-in for a type or universal property *by itself*. It is a concrete particular and, in order to treat it as a sample of a type, one must know what type it is a sample of. But this is exactly what proponents of inner awareness and the phenomenal concepts strategy (including self-representationalism) require: namely, that qualia identify themselves intrinsically. Recall, in this connection, that a key feature of acquaintance with qualia is supposed to be that the mode of presentation is <u>determinate</u>; i.e., the presence of the quale is sufficient to identify it independent of any other description. But to play a determining role it will not be sufficient that my qualia simply look *like that*. There must be some content to "that," and the content cannot be just *whatever that is*, for then it would be indeterminate (cf. Tye, 2009, p.7; Levine, 2001, p.172). All phenomenal concepts would then have the same content: *that x*. If the content of "that" is just a mute pointing, then it is meaningless to say that something looks "like" that. No *likeness* is given in the quale itself, and no basis for deciding in what way the likeness is supposed obtain. Determining what it looks like requires that it be brought under a further concept, which must be applied according to some criteria. And these criteria are precisely the publically regulated deferential concepts that the phenomenal concepts strategy attempts to finesse. In short, to become a phenomenal *concept* (as opposed to just being a phenomenal quality), the experience must be taken as representative. All such taking-as can be mistaken, just as memory can always go awry.[28]

One final point remains to be noted. Certain perceptual demonstratives do involve a kind of non-identifying reference that is immune to error (Campbell, 2002). But the epistemic privilege delivered by simple perceptual demonstratives is referential, not descriptive. In a first-person or subjective frame, one can mentally point to the object *there* ("that x") in a way that cannot go wrong even if the object is elsewhere or non-existent. But this is not the same as *predicating* something of x or *sorting* x into a category (cf. Campbell, 2002, 1999; Pryor, 1999; Geach, 1973). When I judge that I see a "greenish after-image in the left side of my visual field," I can succeed in mentally pointing to "it" even if my judgment is wrong in *every* other way. I can be wrong about whether it is green, whether it is an after-image, whether what I see is in the left side of my visual field, or even whether "it" is there at all.

5 Conclusion: it's just a theory

Yeats is reported to have said "Man can embody truth, but cannot know it." In light of the foregoing, it might be affirmed that we embody qualia, but need not know them. Qualia can still be understood as the "inner marks" of perception, but not the immediate objects of it (a result that also stands contrary to the Transparency Thesis). By cutting the Gordian knot of inner awareness, we can preserve both lived immediacy and a mundanely cognitive form of awareness. But many philosophers resist the line of reasoning suggested here. One worry is that if subjectivity does not deliver foundational knowledge then it cannot play a central role in philosophy of mind and person.[29] Inner awareness, it may be thought, is our methodological and epistemological bulwark against excessively deflationary views. Without robust first-person authority, are we not on the road to eliminativism about consciousness as a whole? I think not. Even though any particular qualitative report may conceivably be wrong, this is no reason to suppose that they are *all* wrong, or that mental life is not really qualitative or subjective after all. True, there is a sense in which the reality of qualitative experience is "just a theory" and not a necessary truth. But considering that virtually all our knowledge has a similar status, this doesn't seem so catastrophic. And it certainly doesn't destroy the traditional idea that experience is "private." But from the fact of first-person privacy has grown a conception of subjectivity as constituted by qualitative "seemings," a conception in which these seemings cannot, strictly speaking, be false.[30] This paper has tried to defuse that conception of subjectivity somewhat. The constructive alternative is to affirm another sort of privacy, in which first-person thoughts can only be shared in one of two general ways, either by transforming their content into a transportable format (conceptual description or external measurement) or by inspiring a similar first-person thought in another subject (empathy and imagination). In either case, something of the original remains unshared: namely, the individual perspective from which it was experienced.

Just as Levine says, the problem of subjectivity is to explain how there comes to be such a thing as a first-person perspective in the world. The strategy pursued here may allow for progress on this problem by distinguishing it from the problem about the ontology of qualia. Explaining subjective consciousness will demand the combined resources of philosophy of mind, embodied cognitive and affective neuroscience, and evolutionary developmental biology. Conceptual and empirical problems remain, including the ontological problem about qualia. But no paradox of subjective duality stands in the way.

Part II
Subjectivity in the Neurobiological Image

5
Interlude: The Neurobiological Image

In the effort to articulate a conceptually adequate and empirically informed analysis of the first-person structure of experience, Part One focused on philosophy of mind and cognitive science. Part Two now shifts to philosophy of biology and embodiment. What emerges is a sort of *neurobiological image* of human subjectivity; a neurophilosophy, broadly construed. In the neurobiological image, we are hide-bound animal subjects, neurologically enabled, ecologically situated, and historically conditioned. It seems to me that this neurobiological image is already out there in the zeitgeist. Increasingly, it guides thinking about human life across the sciences, social sciences, humanities, and the culture at large. In the following chapters I place the neurobiological image in a theoretical framework and discuss its implications.

In speaking of an informal yet guiding "image" of ourselves and our place in the world, I strike a theme from Sellars (1963), who stated that the aim of philosophy is to understand how things in the broadest sense hang together in the broadest sense. Sellars attempted to envision how two very different worldviews "hang together," the *manifest image* of humanity and the *scientific image* of matter and mechanism. To succeed in this synoptic philosophical task, Sellars says, would be to know one's way around in the largest terrain and scheme of things (ibid pp.1–2). As I like to think of it, Sellars' objective is to gain *a sense of where you are*, relative to everything – "...not only cabbages and kings, but also numbers and duties, possibilities and finger snaps, aesthetic experience and death" (p.1). That is fair statement of my philosophical project. I aim to show how subjectivity – a notion drawn from the manifest image – hangs together with the *neurobiological image* of humans, which is a contemporary scientific image.

In Sellars' hands, the manifest image functions as a kind of idealization, an abstraction, a distillation or "reduction" of humanity and its place in the world. It is an image in the sense of an image-schema, a thumbnail sketch of what a person is. Likewise, the neurobiological image is an extract, a synthetic characterization. The critical passage between the scientific and the manifest traverses the gulf from the image of humans as objects (scientific) to the image of them as subjects (manifest). The basic idea of this book, then, is to render an image continuous with life science yet robust enough to be recognizable and even useful to the humanities. Part Two will show that the neurobiological image makes contact with manifest subjectivity in two primary ways: (1) subjectivity is the cognitive, interpretive and historically conditioned achievement a animal coping with its world; (2) the first-person perspective is a variety of navigation, achieved through an affective-cognitive map. It will be seen that the actual phenomenological, psychological, and referential features of the first-person perspective can be given a fine-grained causal/historical explanation by means of cognitive evolutionary developmental biology.

Few philosophers believe that this kind of project can succeed. Dedicated naturalists argue that the lesson of biology is primarily *destructive*: "The manifest image of humankind...takes a major hit at the hands of Darwin's theory, and it is not clear how to maintain the central components of that image" (Flanagan, 2003, p.378). Thoroughgoing naturalists despair of salvaging the idea of the person and its associates. In the Postscript following Chapter 8 I will discuss the destructive consequences of Darwinian naturalism more fully, but in the intervening chapters I suggest that one central component of the manifest image *can* be retained, albeit in a significantly revised form. Persons are, above all, subjects; their experiences embody a point of view. If the scientific image – the new neurobiological version of the scientific image – can really explain *how a subjective point of view arises in the world*, then a central component of the manifest image does indeed survive Darwinism.

Subjectivity is a "central component" of the idea of a person despite being neither necessary nor sufficient for personhood. Being a worldly subject is not sufficient for being a person since elephants and whales and dogs and cats are not persons.[1] And while it may be plausible to suppose that subjectivity is necessary for personhood, nothing much turns on this issue (at least nothing relevant here). There are reasons to think that, e.g. corporations or computers could be persons. If so, then it seems clear that they could be so even in the absence of subjectivity. But in any event, most contemporary philosophers will be inclined to read Flanagan's appeal to "central components" informally, in terms of core or prototypical features rather than necessary and sufficient conditions. Even if it be allowed that some persons (such as corporations) might lack a first-person perspective, these cases are oddities. Prototypically, there is something it-is-like to be a person.

Other central components of the manifest image include such person-relative notions as intention and belief, agency and responsibility, individual and community. These concepts reflect the characteristic logic of the manifest image, in which persons are a fundamental and irreducible component of reality, and the sine qua non of the world as a whole. In short, though Sellars doesn't put it this way, the manifest image is that of traditional humanism. Humanism is the sophistication and refinement of the lifeworld wrought by artistic, religious, and scholarly tradition ("lifeworld" is another term Sellars avoids). By contrast, the scientific image is a worldview organized around the concept of a complex physical system, derived from "postulational theory construction" (19, 25).[2] There is no room in the scientific image for the whole range of person-dependent concepts. Sellars argued that the two schemes are orthogonal to one another, each presenting itself as a complete worldview adequate to the basic furniture of reality.

Since the neurobiological image is the image of the animal as a subject, it differs markedly from anything Sellars envisioned. One example will illustrate the change in the scientific image from the late 1950s to today. Sellars noted that, according to scientific image as he found it, the learned behavior of a mouse cannot literally be said to be *habitual* because "habit" is not a proper scientific concept. It is drawn from the manifest image; it is part of the system of concepts surrounding "person," along with "character" and "action". In particular, habit is a subspecies of intention. Since only persons can act intentionally, only they can act habitually. Thus habit can appear in a scientific explanation only as a metaphor; in good behaviorist form, mice can only be described as exhibiting "habit-like" behavior. But in the contemporary neurobiological image, mice *do* literally have habits. Their habits are both explanatory and explained in the neurobiology of associative learning. Crucially, there is something it-is-like for the mouse to behave habitually. Habitual action is a learned way of navigating the world, and this learning is indeed explained mechanistically.[3]

Can the neurobiological image expand to include subjectivity, and yet remain scientific? Won't this violate the very conceptual schemes in view? A historical sensibility is helpful here. Consider that *motion* itself once appeared to defy the mechanistic worldview. Spinozism was considered confused, not only because of the heretical denial of a transcendent God, but because it ascribed the spiritual powers of motion to the body itself! In fact, one of Descartes' original motivations for substance dualism was his strict adherence to a vision of mechanics which conceived matter as utterly inert. The origin of motion, he reasoned, must come from outside the mechanism since it is unscientific to hold that bodies could literally initiate movement. When Spinoza remarked that we don't know what bodies can do, he meant that movement might one day be explicable in naturalistic terms.

And this sort of historical development in the scientific image is perfectly in tune with Sellars' project. In the early days of the cognitive revolution many psychologists and philosophers wanted to put human meanings back into the science of the mind (Bruner, 1993). They wanted to affirm the reality and scientific necessity of *thought*. Sellars was part of this movement; what he sought was a version of the scientific image enriched in a way that can explain how norms are taken up by an individual person. He suggested that the way to accomplish this revision of the scientific image is to conceive of individual meaning (i.e., norms of mental content) as constituted through communities and language games (Sellars, 1963, p.40). Today, Sellars' program is being carried out by Robert Brandom (2015), among others. But the strategy cannot succeed without *also* enriching the concept of the individual animal and what it brings to the party. It has often been emphasized, with Sellars and all the way back to Hegel, that subjectivity requires intersubjectivity. But at the same time, intersubjectivity requires that there be something like individual subjectivity. This book is a contribution to that side of the project.

6
The Science of Subjectivity: Neurobiology and Evolutionary Development

1 Introduction

There is a venerable tradition in philosophy, in which the conceptual analysis is presented first, with the empirical "detail" to be filled in afterward. In this book, three chapters of philosophy of mind and psychology precede the introduction of relevant biology. But the reality behind this traditional presentation is that good conceptual work is simultaneously informed by the facts. The arguments of the first three chapters were tacitly motivated by what is known about brains, the animals that have them, the environments they live in, and their history. In this chapter I discuss a just a small sample of the abundant and quickly growing research in neuropsychology that inspires this book. As much as possible (it will not always be possible), I postpone discussion of the further theoretical and philosophical implications until the next chapter. First I focus on a kind of thumbnail sketch of the neuropsychological basis of the animal subjectivity.

The neural mechanisms underlying the baseline form of the first-person are among the most well known and intensely studied in all of neuroscience. But because they can operate unconsciously, their importance to subjectivity has not been widely appreciated.[1] Whether or not we are consciously aware of it, there is something it-is-like to cope with the environment, and this activity is at the very heart of subjectivity.

1.1 Coping as affective orienteering

The quintessence of subjectivity is a perspectivally articulated sense of where you are, a set of feelings about the situation, the layout, and where you are headed. We share this elemental kind of subjectivity with a wide array of other animals. Like us, they embody a first-person perspective toward a "salience landscape"(Ramachandran & Oberman, 2007). The landscape is subjectively salient because affect is linked with our perceptual-motor systems via a kind of memory known as the cognitive map. The "coping systems" of percept, affect, memory, and action are very old. Animals have been orienting to their environment for a long time. They have been socializing for a long time. They have been keeping track of their prey and their predators for a long time. The phrase "salience landscape" nicely expresses the idea that the core structure of first-person thought consists in a *concerned engagement* with an environment. Two deeply entwined dimensions can be distinguished. Subjectivity is (1) spatiotemporally perspectival, and (2) affectively engaged. Subjectivity is perspectival in that "I" am specified by the structure of experience itself (see chapter two for more on this point). Moreover, part of the content of first-person experience is *that* it is perspectival (Noe & Thompson, 2004b, p.90, 91fn). Perspectival content concerns possibilities for action from here and now, as well as expected changes that will result. The term "perspectival" is meant to indicate what it is about experience that specifies it as "from here," such that the animal's actions will change the perspectival content in structured ways.[2]

Next, think of the second essential aspect of for-me subjectivity, the affectively engaged aspect, as *where you are coming from*. Subjectivity is the way an individual is engaged in life. It is the basis on which things matter. Where you are coming from is an affectively charged "place" within an affectively structured space, as much as it is a sensed location in a physical space. This kind of subjective engagement with the world is sometimes called *concern*.[3] Thus, the first-person perspective partly consists in an ensemble of proto-phenomenal affective values, valences or "drives" that give structure to the way an animal engages the world. This baseline affective engagement is part of the deep structure of for-me mental states, regulating or governing the first-person perspective.

This affectively structured sense of the local environment is **subjective orientation**. Here is where the phenomenology of the first-person makes basic contact with ecology. Baseline subjective orientation does not require a self that is specifically conceptualized *as* an "I". Nor does it require full-blown reflective consciousness or metaconscious awareness of itself specifically as qualitative or internal content. It need only enact a first-person point of view. Baseline subjective orientation, in this sense, is achieved by most vertebrates (at the very least), including those that may lack reflective abilities. Finally, although it is often unconscious, baseline subjectivity does not take a classically psychoanalytic form so much as an ecological form. The neurobiology underlying this ecological first-person perspective will be the focus of the following discussion.

1.2 Evolutionary development as a research program

Which species achieve a baseline form of first-person subjectivity? This is an empirical question. A framework offering resources for investigating this empirical question can be found in *evolutionary developmental biology* (evo-devo), specifically *cognitive* evo-devo. Cognitive evo-devo is a research program that aims to study cognitive evolution indirectly, through neural trait polarity. The evolution of the brain is the primary locus of the evolution of the mind,[4] and earlier forms of the brain existed in animals that enacted earlier forms of subjectivity. Past brains have been genetically and developmentally conserved in ways that function to constrain subsequent evolutionary elaborations. So, historically primitive cognitive processes can be tentatively identified by tracing the evolutionary development of the brain across species and phyla (Jacobs, 2012). This historical and comparative method contrasts with much of today's so-called "evolutionary psychology," in which cognitive function is identified first and selection for that encapsulated capability is subsequently inferred. Full discussion of *homology thinking* (Ereshefsky 2012) and its distinction from mainstream evolutionary psychology is reserved for chapter seven.

Two key themes of evo-devo research are neural plasticity and developmental covariation (Charvet & Finlay, 2012). An initial insight is that in complex and changing environments, neural plasticity (and along with it, delayed development within the individual lifespan) has been selected for. This means that *learning ability* is a cognitively primitive trait relative to many observed functions of adult mammalian brain areas, which are epigenetically specified in concert with neural development and reorganization. In general, selection does not operate directly on neocortical regions, but rather on whole animals, growing and developing in their environment. For example, with respect to social cognition, evo-devo considerations show that "the evidence supporting *selective* changes in isocortex or brain size for *isolated* ability to manage social relationships is poor" (Charvet & Finlay, 2012, p.71). Instead, a flexible brain with a long developmental trajectory is adaptive in highly variable social ecologies. Further, the neocortical plasticity that embodies our ability to learn was selected and developed *from a previously existing trait in the population.*

Before the development of the neocortex, there were analogues of this coping ability that drove subsequent evolution. These "primitive" traits (i.e., historically prior) have been conserved, and can be identified by looking at patterns of covariation across species. What is this ancient form of individual coping? In a word, it is *navigation*.

Explaining how animals navigate is among the fundamental tasks of biology. Its explanation is framed in terms of the evolution of cognitive maps, and this is also the framework in which to pursue an explanation of the first-person perspective. To cope is to *find a way from here*, and this is a kind of navigation. As philosophers since at least Kant have often pointed out, subjectivity is spatiotemporal sensing. But cognitive mapping is a far cry from a Kantian "transcendental aesthetic." It is a dynamic memory process that coordinates spatiotemporal position with affect and action.[5] When placed within an evo-devo framework, the upshot of this idea is that first-person subjectivity is the expression of a highly conserved neurogenetic structure that evolved from prior navigation systems dating as far back as the Cambrian explosion, as many as 500 million years ago.

So, the science of subjectivity in view here is couched in an evo-devo framework, taking navigation and the cognitive map, respectively, as the model problem and explanation. These choices of framework and paradigm may be compared with the course taken by Evan Thompson in his neurophenomenological opus *Mind and Life* (2007). Thompson, too, grounds the philosophy of subjectivity in biology. But, rather than looking for the specific evolutionary history of first-person thought, Thompson finds a more general analogy between the emergence of life and the enaction of mind. Accordingly, his model phenomena are not navigation and the cognitive map but rather autopoeiesis and the self-organizing system. The philosophy of biology in which Thompson grounds his philosophy of consciousness is emergentist, denying that "objective," mechanistic or naturalistic biology can explain vital subjectivity.[6] By contrast, I think that these are the *only* things that can explain it. As a caveat, I stress that the empirical theories discussed below are provisional. As research continues at its currently feverish pace, the models will surely be modified and complicated, and possibly thrown out altogether. Further, very many aspects of subjectivity are omitted from this chapter, including especially our social orienteering, i.e. intersubjectivity. So, do not mistake this discussion for an attempt at a comprehensive final account! But if the general direction is sound, then something like the following will turn out to be the right account of baseline subjectivity.

1.3 Limbic, limbic, limbic

Roughly, the neural bases of our ability to take up a perspective on the world lie in and around the *limbic system*. For simplicity I try to use that phrase as much as possible, but there is significant terminological variance in the literature and sometimes I will speak of an "olfactory-limbic system" or "mesolimbic / dopaminergic system" (ML-DA). I should also clarify at the outset that talk of *the* system is a simplification, since there are many interrelated and overlapping mechanisms linked with one another.

In the literature on consciousness it is canonical to suppose that limbic processes are deeply unconscious and irrelevant to subjectivity. For example, Giulio Tononi's beautiful book *Phi* expresses the mainstream view that consciousness is exclusively a matter of corticothalamic connectivity, while the limbic functions of cognitive mapping and affective regulation are merely the "slaves" of this higher system (Tononi, 2012, pp. 72–5). But when subjectivity is considered as the first-person perspective (rather than as inner awareness, see Chapter 4), it becomes apparent that the canonical view is misplaced.

The phrase "limbic system" is informal. It loosely refers to a set of interconnected anatomical structures at the base of the cerebral cortex, deep within the temporal lobes. The word "limbic" describes the layout of many of these structures, which wrap around the tip of the brainstem, forming a sort of cap between the stalk and neocortex. However, the limbic system also includes some brainstem nuclei that secrete neuromodulators diffusely across the brain. The hippocampus is the hub of the limbic system. It is centrally located on the inner surface of each hemisphere and has a long phylogeny, traceable back to early vertebrates. The entorhinal cortex is a distinct anatomical and functional subregion nearby, in the parahippocampal gyrus. It receives and integrates feedback from all the primary sensory cortices, and it is strongly interconnected with the hippocampus. Other closely connected anatomical regions, important for the discussion below but not always considered part of the limbic system, include the olfactory bulb and ventral striatum.

Attitudes toward "the limbic system" as a unified theoretical concept have varied considerably through the latter half of the 20th century. There is still some debate about the value of the term, but it is widely used as a general placeholder that stands in broad contrast with "neocortical" and "brainstem" regions (though limbic functions clearly involve at least some structures in both of the other two). MacLean (1952) popularized the idea that there is a functional and anatomical unity to the limbic system, and MacLean (1960) went on to make the limbic system the keystone in his "triune brain" hypothesis, dividing gross brain anatomy among the Big Three. MacLean essentially argued that there is not one brain but three, each layer embodying a relatively independent and phylogenetically conserved neural system. In this respect, MacLean was a forerunner of the evo-devo approach. But the triune brain hypothesis can be dangerously misleading because it vastly oversimplifies the degree of connectivity and functional interdependence across brain regions. Gross anatomy does not cleanly encapsulate functional organization, and the triune structure MacLean pointed to does not accurately reflect phylogeny in the way he supposed. Further, there was no "devo" in Maclean's version of the theory; no developmental story. The triune brain is simply the adult gross morphology, genetically determined.

So the triune brain hypothesis, of which the concept of the limbic system was a part, was an important but flawed precursor of evolutionary developmental neurobiology. A few caveats are therefore in order. When psychological function is in view, the limbic system is sometimes called the *emotional brain*, but this may be misleading since the hippocampus is also the anchor of memory. Phylogenetically, the limbic system has been called the *old-mammalian brain*. This is also misleading because it now appears that the common ancestor of reptiles, birds, and mammals had a well-articulated limbic structure (the medial pallium), and that some homologous tissue (the olfactory glomerulus) is present in all animal phyla including the invertebrates (Jacobs, 2012). Anatomically, the limbic system is sometimes called the "chemical-visceral system" since it directly processes olfactory and endocrine information. In this respect olfaction can be thought of as the external immune system, in which chemistry in the environment interacts more or less directly with the internal milieu. But again, this can be misleading since there are many robust connections between all the sensory cortices, the thalamic nuclei, and the limbic system.

The following brain structures are commonly considered to be part of the limbic system: hippocampus and parahippocampal gyrus including the entorhinal cortex, hypothalamus and mammillary bodies, amygdala, cingulate gyrus, and fornix. There are also integral connections with the striatum/nucleus accumbens, ventral tegmental area, and other prefrontal cortices. So, whether a given structure is or is not part of the limbic system can be indefinite. For example, the amygdala evolved somewhat later than many of the other organs on the list, and is more closely associated with thalamus and neocortex. But due to its important role in emotion it is usually considered as part of the limbic system.

Different researchers use the limbic idea in slightly different ways to talk about a variety of overlapping structures and functions. Jacobs (2012) uses "olfactory limbic system" to refer to networked structures distributed across the hippocampus and its subfields, olfactory bulb, olfactory cortex including piriform gyrus, entorhinal cortex and amygdala, and the septal nuclei.

With the above in place, I can now turn to describe some of the science of subjectivity. Here is a brief sketch of the story that follows: The function of baseline subjectivity is learning to cope. Coping is a kind of *navigating*. Learning to cope is explained by *mapping with affective valence*.[7] Mapping with affective valence emerged first in OB and then in the olfactory limbic system. In their contemporary forms, these systems display delayed development, adult neurogenesis, and large relative size across phyla. Thus, the explanation of baseline subjectivity considered as the first-person perspective is to be found in the ensemble of limbic mechanisms, embodying a cognitive "trait" that is phylogenetically primitive in relation to conscious awareness.[8]

2 The Olfactory Spatial Hypothesis

Roughly, the *Olfactory Spatial Hypothesis* (OS) is that navigation was the primary selected function of olfaction, while fine odor discrimination is secondary (Jacobs, 2012). The prevailing assumption has always been the reverse, namely, that acuity in odor discrimination is selected for and that spatial orientation is a separate cognitive ability.[9] But Lucia Jacobs (2012) argues for a reinterpretation along broadly evo-devo lines. For several reasons, the OS is worth presenting in some detail. First, it represents a fascinating and robust area of research into the neural basis of the first-person perspective that has not been generally recognized as relevant to the science of subjectivity. Second, it illustrates the importance of an evolutionary perspective on cognitive neuroscience. In the absence of the evo-devo framework, it is difficult even to frame the right questions about subjectivity.[10] Typically, the attempt to explain or naturalize subjectivity or consciousness is framed analytically and ahistorically, by asking questions like "What computational or representational model of conscious awareness can be constructed?" "What is the conceptual or epistemic role of subjectivity?" These are legitimate questions, of course. But they do not connect in any interesting way with the biological science. The evo-devo framework guides investigation by focusing attention on the "archeology" of subjectivity – the historical basis of the phenomenology of the first-person. Questions that emerge include "From what did the first-person perspective evolve?" "What is the history of our present forms of consciousness?" "How did animals like us become subjects?"

To clarify, I am not particularly committed to Jacobs' specific claim that the navigation function preceded fine discrimination. I am more interested in the fact that navigation (orientation & engagement) is the fundamental cognitive achievement on which subjectivity is based. The neural mechanisms with which contemporary species like us cope are the "descendants" of the neural mechanisms with which our ancestors coped with theirs. Jacobs argues that olfaction (and navigation more broadly) has been misunderstood, and that progress can be made by adopting a theoretical perspective that is more ecologically valid and pays careful attention to evolutionary development. Both methodologically and substantively, then, the OS hypothesis is an important contribution to the science of the subjectivity.

Jacobs argues for OS in several ways. First, the OS hypothesis makes the best sense of the neuroanatomy of olfaction, which remains mysterious on the prevailing view. If the selected function of olfaction is fine odor discrimination, then the relative size of the olfactory bulb (OB) should correlate most directly with discriminative power. But it does not. In addition, Finlay, Darlington, et al (2007) documented that relative OB size (and along with it the relative size of the entire olfactory limbic system including hippocampus and amygdala) varies independently of the pattern of scale for the rest of the brain. Since fine odor discrimination does not account for this observation, another explanation is required. Jacobs' Olfactory Spatial Hypothesis offers one: The pattern of scaling in the vertebrate OB reflects directional selection to use information about chemical distributions in the environment for spatial navigation (Jacobs, 2012, p.10693). OS therefore predicts that olfactory bulb size reflects navigational demand. That is, ecological factors relevant to the navigation task should sort OB size according to an animal's niche, together with other developmental and cognitive factors.

The OS prediction fares well in finding the order in the otherwise noisy variance in OB development across species. In making her case, Jacobs draws an ecological distinction between "detector" and "predictor" species. Detectors are foragers whose food is either easy to find (grass) or very difficult to find (clouds of insects). These species don't need to think very hard about where their food is. By contrast, predictors are animals that find prey (or predators) moving through the environment in ways that can be tracked with the right information. Predictor species, then, invest in spatial mapping systems to cope with this problem. Armed with the ecological distinction between detectors and predictors, Jacobs shows that the pattern of relative OB size does indeed co-vary with the navigation demand faced by each kind of animal.[11] She concludes that olfaction evolved as a specialization for mapping spatiotemporal stimuli into functional associative memory structures. She further argues that this organization is embodied in a parallel map structure (Jacobs, 2012, p.10698). I pause to introduce this last idea.

2.1 Parallel Map Theory

Jacobs' Olfactory Spatial Hypothesis is a complement to her earlier work on the Parallel Map Theory. The OB is not usually understood as a spatial mapper. Instead, it is well established that mammalian cognitive maps exist in hippocampal networks of associative memory. But hippocampal architecture may have evolved from prior olfactory maps. Jacobs applies her prior (co-authored) functional model of how the hippocampus performs its navigational function to the OB, and argues that a similar mapping mechanism may exist there (Jacobs LF, Schenk F, 2003). According to Parallel Map Theory (PMT), animals navigate by means of the coordination of several hippocampal subregions, each tracking and integrating different sorts of information. The very notion of a cognitive map incorporates two elements, that of a topographical map and that of a compass or relative heading map.[12] First-person spatial orientation is made possible by the coordination of these two kinds of information and PMT is one current model of the neural mechanisms that underlie this ability.

From within an evo-devo framework, Jacobs & Schenk (2003) argue that these mapping mechanisms have undergone several evolutionary stages. The first mapping stage is a *Bearing Map*. This is constructed as the animal explores the environment by moving along gradients of stimuli that provide directional information. The bearing map is a compass that allows the animal to predict future position within the flow of graded information, even in the absence of a topographic (topo) map, and even in a previously unexplored space. The bearing information specifies "nearer/further," "older/newer," "downwind/upwind," etc. In mammals, it is thought that the neural substrate of the bearing map is the hippocampal subfield known as the dentate gyrus. A second mapping stage is the *Sketch Map*. The sketch map is the topological memory of local landmarks, using sets of remembered positional cues in the bearing map to derive representations of discrete objects in spatial and temporal order. In mammals, the proposed neural substrate for the sketch map is area CA1 in the hippocampal structure known as Ammon's horn. According to the Parallel Map Theory, information from these two "parallel" maps is brought together in a third *Integrated Map* that recodes topological information onto the bearing space. In mammals, the proposed neural mechanism underlying the integration of this information is thought to be in hippocampal subfield CA3. The integrated map makes available information analogous to a pointer on a GPS. An integrated parallel map mechanism of this kind allows the navigator to find shortcuts "from here" through a remembered environmental layout.[13]

The close relation between the olfactory system and the hippocampus is well known.[14] Hippocampal cognitive maps are learned from sensory experience (and affective feedback – more on which below). In contemporary and sophisticated animals like us, multisensory information descending from the neocortex converges in the entorhinal cortex and interacts with the various hippocampal subfields. But the olfactory system is older than the neocortical perceptual systems. Indeed, the olfactory bulb is older than the hippocampus itself.

On the OS hypothesis, olfaction is not conceived of simply as "smell." It is understood as the ongoing integration of chemical and vibratory sensory systems, coordinated with one another in the activity of exploring the environment. Olfaction is fundamentally different from (and simpler than) the other senses in that it requires no transduction into the neuropharmacological system. Chemicals at the periphery directly interact with neural chemistry without having to be converted from other forms of energy. This is why it was noted above that the olfactory system is sometimes described as an "external immune system," responding to evanescent and turbulent information about the past, present, and future of the animal (Bargmann, 2006). The analogy to the immune system and the homeostatic self-regulating activity of the animal is apt, pointing toward the deep alliance between perception, affection, and memory in subjective mental life.

Jacobs argues that olfaction has other peculiar features that lend credence to the idea that its primitive function is navigation, and that it is an earlier version of the parallel map mechanism. Two universal features of olfaction, Jacobs argues, can be exploited for navigation. (1) Changes in the intensity of a single stimulus can produce qualitative changes in perceived odor, i.e., low and high concentrations of a single chemical can produce perceptual judgments that the samples are qualitatively dissimilar (e.g., orange ->lemon). This shows that there is a structured "olfactory space" that has neighborhoods or attractor basins. (2) Perception of mixed plumes depends on prior learning. Subjects of all types, from human to lobster, demonstrate the ability to switch between perceiving odorant mixtures as either synthetic wholes or as the dual presence of two elemental ingredients. This ability to perceive "odor objects" shows that unique chemical ratios can become distinctive landmarks within the graded olfactory space. From these observations, Jacobs shows that it is possible to construct a metrical representation of the spatial relations among odorants. Thus, the state space of neural networks in the OB can become a cognitive map of the local environment. Jacobs further argues that these two universal features of olfaction are evolutionarily prior analogs of the parallel maps found in the hippocampus.

The evo-devo punch line of the OS hypothesis, then, is that olfactory navigation is *the* phylogenetically primitive cognitive ability. Our kind of brain is descended from the olfactory architecture that first evolved to map chemical distributions in an aqueous environment. Within the constraints of this existing olfactory system, expansion into new niches through the discovery of new kinds of information brought with it newer, more sophisticated ways of coping – visual, auditory, and finally linguistic (See also Sarnet, HB, Netsky MG, 1981). If this is right, neocortical abilities would retain "…the primacy of olfaction, i.e., olfactory guided navigation, as the ancestral function of the forebrain, and they would for this reason eventually converge on a similar neuroarchitecture" drawing on overlapping neural resources (Jacobs (2012), p.10698). The idea that ancient systems are retained in contemporary elaborations is a distinctive characteristic of the evo-devo approach. Surprisingly, then, the deep structure of human subjectivity is to be explained by olfaction – which evolved before vision and before language.

2.2 Of limbic loops and cognitive maps

Jacobs is by no means the only theorist to present a model relevant to the first-person perspective and based in the limbic system. Like Jacobs, Walter Freeman (2001) holds that a dynamic entorhinal-hippocampal system was the original basis of vertebrate experience, and that other sensory modalities were later integrated with this persisting older system. He calls this dynamic complex the *space-time loop*, and argues that it is the functional core of the limbic system and the "principle architect of action in space-time" (Freeman, 2001, p.35). Freeman argues that subjective space and time are "lived" through a dynamic interaction of hippocampal preafferent activity and multisensory feedback convergence in the entorhinal cortex. In some respects, Freeman goes further than Jacobs, arguing that the first-person is not based on "archival maps" or static, snapshot, bird's eye (third-person) views located at addresses in a memory bank (Freeman, 2001, p.104). He emphasizes that cognitive maps (which are at the core of the OS hypothesis) are not to be understood as "internal world models" or virtual realities *in which* subjective experience takes place. The animal's first-person perspective does not correlate directly with minimally identifiable receptive-field activations in any particular network.

Arguably, the Parallel Map Theory's "integrated map" idea, described above, conflicts with Freeman's claim that subjective orientation is a distributed property of the system. On Freeman's model, navigation is governed by differential changes in high-order patterns of neural activation that are coupled with one another in the context of an embodied animal in a real environment.[15] While it seems Jacobs would certainly agree with all that, her PMT arguably embraces more "cognitivist" assumptions about internal representation. In one sense (most important), this is an empirical question: either there is or is not an integrated map in CA3. But there is also a conceptual point here. Whatever information turns out to be in the head, its processing remains situated in a body and in an environment. The first-person perspective is that of an animal, not that of a neural network. This is a truism, but even as it is over-emphasized by some proponents of embodiment, it is lost on some internalists.[16] Keeping this firmly in mind, it is clear that there is a high degree of coherence between the OS hypothesis (particularly the idea of the bearing map) and the Limbic Loop model. The result in both cases is an agent-level form of cognition in which "space" is just the explored and explorable layout, while "time" is the practical lapse of exploratory actions.[17] What is important here is that both PMT and Limbic Loop models begin to explain *how there is a first-person perspective*; i.e., how experience comes to be ecologically indexed to an agent-level form of navigation "from here" that must be actively maintained over time and enabling practical action and expectation for-me.

More recently, Freeman (2008) has expanded his view in a way that connects it quite directly with the approach taken here:

> In order to track a sequence of molecules toward a potential source of food or away from a dangerous sink in someone else's maw, an animal had to develop the capacities to navigate in space and time, and to predict the directions, distances and travel times from the present site to significant locations. These neural capacities are apparent in the organization of even the simplest vertebrate forebrain (Herrick, 1948), that of the tiger salamander (Figure 1). Each of the two hemispheres consists of an anterior third devoted to sensation, mostly olfactory, a

lateral third devoted to motor functions, and a medial third devoted to the capacities that are required for the spatiotemporal guidance of movements. This medial third is the recognizable forerunner of the hippocampus, which in vertebrates ranging through reptiles and mammals to humans is recognized as being essential to form episodic memories and remember places, that is, for temporal orientation, and for orientation of action in space. This capacity was designated by O'Keefe and Nadel (1978) as a "cognitive map", after the concept developed by Tolman (1948), but psychologists and roboticists have concluded that this concept is merely a metaphor. They propose that animals (Jacobs 1994) and robots (Hendriks-Jansen 1996) do not literally have maps in their heads, but instead they have the capacity for site-specific regulation of their behavior in spacetime. (Freeman, 2008, pp.4–5)

This passage contains several key themes. First, the reasoning is conducted according to the "homology thinking" characteristic of evo-devo explanation (Ereshefsky, 2012). Second, the ecological function of the first-person perspective is conceived as navigation and orientation in the local environment in real time. Third, navigation is explained as a kind of mapping, mediated by memory and expectation. Fourth, on Freeman's interpretation, the terminology of mapping is metaphorical, referring only to a "capacity for site-specific regulation" of behavior. My immediate concern is with the last of these three ideas; discussion of the other two will continue throughout the next chapter.

Freeman's claim that the cognitive map is a mere metaphor reflects a larger anti-cognitivist theme in Freeman's work, one that is congenial to embodied phenomenology or *enactivism*.[18] What is the difference between literally having a map and having a capacity for site-specific regulation of behavior? In the end, the difference comes down to a difference in the interpretation of dynamic neural processes. Freeman is concerned to clarify that the cognitive map is not a static "snapshot" of the environment (it is not file footage stored as images), and it is not encoded like a street map, in a bird's eye view. That is, the cognitive map is not a simple *topographical* code for the external space (cf. Grush, 2006). Instead, hippocampal place neurons are activated by more minimal configural cues that are used to govern the global process of navigation. This ongoing activity of "orienteering" is undertaken by the whole animal, involving the constant construction and reconstruction of meaning through engagement with the world. Freeman is concerned to clarify that the whole of this meaning-construction is not simply "read off" the hippocampal place cells in episodic memory. So much is certainly true. And it is relevant because it helps to point to the specifically *first-person* nature of animal navigation (that is, if the map could be simply "read" from outside, this would indicate a third-person or objective form of representation). Nevertheless, none of this makes the language of cognitive maps into a *mere* metaphor. Possibly, there is an analogy in play somewhere, with a comparison to a street map. But the mapping mechanisms are real and they literally perform a mapping function. They encode spatiotemporal information, gained through exploration, guiding and orienting current activity. That's a map, literally. Exactly what form the maps take, exactly how they encode information, their precise degree of similarity to street maps, is strictly an empirical question. But to retreat to talk of a "site-specific capacity" is to mystify the process, and to lose the explanatory value of the idea. Cognitive maps are mechanisms, not capacities.

If hippocampal maps are not topographical maps or images, then what are they? According to PMT, they are of two kinds: gradient maps and topological maps (Jacobs, 1996). In gradient maps, samples are ordered by intensity along dimensions of value. For example, an olfactory gradient is sampled in a series of sniffs, and this can be the basis for construction of a continuous vector where "distance" is a function of intensity and time. This is a so-called "Euclidean" value code, allowing the animal to move up or down the gradient. The topological map then plots temporary stimulus objects onto the more stable dimensions defined by the gradient map, allowing for inferences about transitivity, temporal ordering, and novel "short cuts." The two parallel maps are not "topographic" in the sense that they are not *isomorphic with the space that they map*. That is, the neural activation will not *look like* a map. But their integration in working memory guides exploration and governs relational learning (Jacobs, 2012, 1996). This is just the function performed by external maps and compasses in orienteering/navigation.

But Freeman's anti-map polemic is not mere crankiness. Like Panksepp (see below), he is pointing out that canonical ways of thinking in cognitive and behavioral neuroscience have been blind to the first-person, subjective character of limbic function. After all, the hippocampus is probably the single *most* studied and best understood neural structure! But background assumptions about the sophistication, privilege, and uniqueness of human consciousness, together with the theoretical alliance between behavioral neuroscience and the cognitive conception of the unconscious (see Chapter 3), have prevented us from seeing that animal navigation and orientation is the ecological function of baseline subjectivity.

So far I have been discussing work that most directly models the *perspectival* dimension of baseline subjectivity, identified at the outset of this chapter. I next turn to the other fundamental dimension of for-me thought, concerned engagement.

3 Primary affective modes

I use the phrase "primary affective mode" to refer to a central theoretical concept in affective neuroscience, though this precise phrase is not used in any very systematic or rigorous way by its proponents. Affective neuroscience is a research program led by Jaak Panksepp and his colleagues (2012, '10, '07 '06, 1998). Panksepp's synoptic work *Affective Neuroscience: The foundations of human and animal emotions* (1998) offers a rebuilt and reorganized approach to the neurobiology of emotion and affect, presented under Panksepp's distinctive rubric. The fundamental premise of affective neuroscience is that our most basic experiences arise not from the unique human forebrain but from subcortical, midbrain, and limbic processes shared with many species. Panksepp argues that there is a level of "affective consciousness" that animates and pervades our whole mental life in ways that are typically unrecognized by researchers. The stated aim of affective neuroscience is to identify and model the primary affective modes and their functions in the "MindBrain." So far, seven such modes have been experimentally documented: SEEKING, RAGE, FEAR, LUST, CARE, GRIEF, and PLAY (Panksepp, 2012, pp10–11).

Despite what readers coming from psychology, philosophy, and cognitive science may generally assume, the proper contrast to "affective neuroscience" is not "cognitive neuroscience" but *behavioral neuroscience*. Behavioral neuroscience is the huge and thriving subfield of biology dedicated to the genetic, developmental, and neural bases of behavior. Explanations in behavioral neuroscience typically try to avoid mentalistic terms and ideas, preferring to stick with directly manipulable and observable variables. In a sense, behavioral neuroscience is the true descendent of behaviorism in psychology. When the cognitive revolution dislodged behaviorism from psychology departments across the U.S., students of behaviorism did not simply disappear. They moved to the biology department.

In behavioral neuroscience there is a methodological imperative to avoid anthropomorphism. This is sometimes justified on grounds of parsimony, but more often it simply reflects a general commitment to good science. There is a belief that hard science, unlike social science and the humanities, cannot theorize about the mind and consciousness. Subjectivity can be left to humanists, critics, philosophers, and other mystics. Over the last 25 years, an alliance with cognitive psychology has flourished as behavioral neuroscientists have helped themselves to the seemingly safe and respectable notion of the *cognitive unconscious* – the idea that any information process not introspectively explicit is not a for-me process, and can therefore be treated completely apart from the study of experience (see Chapter 3). Alcaro, Huber, & Panksepp (2007) refer to this unholy alliance with cognitive science as "Neurocognitive behaviorism" and contrast it with their own "active-organism view," based on the experience of affect. Concerned engagement with the environment, they maintain, is rooted in an "internally generated action tendency" standing at the heart of biological cognition (Alcaro, Huber, & Panksepp, 2007, pp.292–295). The point of the authors' emphasis on internally generated affect is to distinguish their approach from behavioral learning approaches based on conditioning. The generator of the internal "urges" is held to be the mesolimbic-dopaminergic system (ML-DA), which both animates neocortical thought and can also function on its own, without neocortical input. This system, they argue, has been significantly misunderstood and underestimated in behavioral neuroscience, despite being the locus of intense research.

The central hypothesis is that there are global control systems that give experience to its tone and dynamics to experience. These are the *primary affective modes*.[19] The externally observable aspect of an affective mode is an open-ended behavioral trope or scheme, the precise content of which is specified within the animal's situation. But Panksepp maintains that these affective modes also have a raw feel or subjective valence. Importantly, primary affect plays the role of the unconditioned response in behaviorist learning regimes. All associative learning must draw on *unlearned* values and valences if they are to get off the ground. The theory of primary affective modes elaborates on the notion of the unconditioned response, transforming it from a simple positive or negative signal into an ordered and subjectively salient pattern of spontaneous activity. And this general idea is not new to neuroscience. As far back as the 1930s it was shown that, when the feline hypothalamus is stimulated, cats will display a pattern of behavior that looks for all the world like rage. This behavioral attitude/style is a complex whole, but it can be evoked with a simple manipulation at a single site.[20] Panksepp long ago embraced the idea that subjective quality is intrinsic to the neural mechanisms of affect.

SEEKING is a good example of a primary affective mode, and it has gradually become a centerpiece of this research program. The idea is that mammals are motivated to forage and explore (i.e., navigate, orient, engage) by an open-ended *curiosity system* governed by networks running through the lateral hypothalamus (LH), the ventral tegmental area (VTA), and the ventral and dorsal striatum (VS and DS, also called the nucleus accumbens). When LH is electrically stimulated, animals always engage in a sustained exploratory pattern (Panksepp, 1998, pp.151–3). SEEKING activity is not strictly "stimulus-bound," focused on obtaining some particular consummatory reward. It is a sort of *generally interested* frame of mind. It is a motivated but nonspecific mode of action that serves as its own reward. It is a Pankseppian adage that *reward is exploration* (Ikemoto, 2010b).

In this framework, ecologically important places are learned (mapped, remembered) thanks to the enthusiasm of the SEEKING system. Here the theory of the primary affective mode parallels, in a certain respect, the OS Hypothesis. Where the OS Hypothesis reverses the functional priority of navigation and fine discrimination, primary affective modes reverse the priority of seeking and "liking" or reward. This comes out most clearly in the way the SEEKING system is tied to phenomenon of *place-preference* (along with another primary mode, CARE). Place-preference has been extensively studied in behavioral neuroscience.[21] Animals can acquire positive associations, expressed in preferential behavior, for particular places in their world. In many studies, place-preference is induced chemically by injecting dopamine or oxytocin directly into the "reward centers" of the brain, principally the hypothalamus. Place-preference is typically understood as a side effect of the animal's "liking" for the drug. That is, in the behaviorist framework, the neural system has an appetite for the neurotransmitter, learns to associate the place with obtaining the drug reward, and so goes there in order to consume it. But Panksepp reverses this thinking by holding that the foraging activity itself is what is important, not the "consummatory behavior." When the foraging impulse arises, animals begin to explore. And after a period of roving, animals eventually turn toward *home*. Home is the prime example of a preferred place, a place the animal returns to again and again. But home is where the ML-DA system deactivates, not where it engages. Affective neuroscientists therefore argue that place-preference is not explained by a simple craving for dopamine. Instead, when mammals reach home, the SEEKING system deactivates and foraging activity comes to a halt.

The ML-DA system underlies an intrinsic desire to explore and engage, and both the activation and deactivation of the system can be experienced as "rewarding" depending on context. Behavioral learning in ecological context, then, is not to be understood simply as appetitive/ consummatory behavior driven by dopamine. Rather, it is a structured whole, the function of which is to learn about and engage with the world of concern. So it seems that country singers are not the only ones with rambling fever, and they are not the only ones to feel the tug of home.

One important set of experiments was conducted as follows (Mendelson, J, 1972). One side of a test area was strewn with small objects while the other side was kept clear. ESB was applied to lateral hypothalamus only when the mice were in the cluttered side of the area. Soon the mice began systematically to carry flotsam & jetsam to the empty side of the area, where the ESB was turned off and they dropped it. Panksepp hypothesized that the ESB stimulation had triggered a state of mind that normally (ecologically) occurs in SEEK mode, when the animal is foraging. Upon reaching home, the foraging impulse naturally declines and they put the found objects down. This is what was simulated by extinguishing the ESB on that side of the test area (Panksepp, 1998, p.155).

3.1 Assessing affective neuroscience: evidence of unconscious subjectivity

The data Panksepp appeals to are widely accepted, replicated, and elaborated. But his interpretation is not. His many proclamations that primary affective modes are *conscious* have mostly fallen on deaf ears. There are several reasons for this. One has been mentioned above – the lingering behaviorist attitudes in neuroscience. Panksepp makes much of this sociological fact, and suggests that affective neuroscience has been marginalized primarily for this reason. But there is more to it. First, affective neuroscience is consistently presented as a form of dual aspect monism. That grand metaphysical scheme is not easily understood, and in any case it adds nothing whatever to empirical explanation. Panksepp's repeated emphasis of it seems misplaced, and most of his readers are put off by it – whether they are biologists, psychologists, or philosophers.[22]

But much more important to the reception of affective neuroscience than either institutionalized behaviorism or suspicion of Spinozist heresy, there is the almost self-evident fact that these affective systems can and do operate unconsciously! While there is good empirical basis to accept that there are primary affective modes of the sort Panksepp proposes, it is abundantly clear that we are very often *not* aware of them, their operation, or of which mode we may be in at the moment. We are not aware of our valenced associations or our underlying drives and motivations in any straightforward sense. And when we do become aware of (able to report) a mood or feeling tone, such as the "intense interest" said to characterize SEEKING, this introspective awareness is *always* accompanied by top-down, more prototypically cognitive neocortical activations. Since we humans aren't necessarily conscious of our primary affective modes, it seems more than a bit perverse to insist that mice are. So affective neuroscience seems by turns unhelpful (because incompatible with physicalism) and misguided (because contrary to clear experience).

In arguing that the primary affective modes are conscious, Panksepp employs a neurobiologically updated version of Mill's argument for the existence of other minds, known as the Argument by Analogy. Panksepp's version runs as follows. We know that human midbrain and brainstem structures are essential to affect, and we know that these structures are strongly homologous across mammalian species. When humans are stimulated in these brain areas they report conscious affect and they produce behaviors that are similar to those observed in model animals under corresponding conditions. Therefore it is highly probable that the model animals are having conscious experiences. Call this the *argument from neural homology*.[23]

There are several things that are right about the argument from neural homology. After all, it is this reasoning that guides most all research in behavioral neuroscience. There would be little point in using animal models to study behavior if neural homology were not a basis for comparison (cf. Love 2007). Panksepp suggests that the refusal to extend this absolutely standard reasoning to the case of consciousness is arbitrary. And certainly this argument accords well with the evo-devo framework, a key feature of which is *Homology Thinking* (next chapter). The most compelling aspect of Panksepp's argument from neural homology is its use of the comparative method, grounded in the continuity of species.

Also note that Panksepp is not completely alone in extending this standard reasoning to subjective experience. Martha Farah (2010) takes up the argument from neural homology in a neuroethics context, though she employs a more standard ontology than Panksepp's dual-aspect monism. She argues that our best current neuropsychology, together with our best philosophical accounts of supervenience, render Mill's analogy stronger than ever. It is a mainstream position in philosophy of mind that conscious states "supervene" on neural states. *Supervenience* is a relation between first-order and second-order properties, wherein any two objects with the same constellation of first-order (physical) properties necessarily have the same higher-order, "supervening" properties. In this case, the first-order properties are neural structures and processes, and the second-order properties are conscious experiences. Thus, Farah argues as follows:

- The human ability to suffer supervenes in important ways on the anterior cingulate cortex (ACC).
- Other mammals have very much the same ACC humans have.
- Therefore, those animals suffer in very much the same ways humans do.[24]

The force of this argument depends, in part, on just *how much* homology is required, or what is the same, on *which* homology is relevant. Obviously, the whole of the neocortex is not homologous across mammalian species in the way that basal forebrain and brainstem structures are. The mainstream view is precisely that these neocortical differences are what make the difference for consciousness. By contrast, affective neuroscientists and neuroethicists like Farah hold that the "higher" neocortical cognitive processes are necessary only for the conceptualization and report of the basic affective consciousness that is shared with nonlinguistic species.

The problem underlying this disagreement is one that was encountered earlier, in Chapter 3. It is the *problem of unreportable experience*. The difference in this case is that the putative animal experience is unreportable, not because the subject has suffered brain injury or because of a clever stimulus presentation, but simply because the subject is a nonlinguistic creature to begin with. But this difference doesn't change my approach, and I take the same conclusion here that I drew earlier. The preponderance of data collected by Panksepp and many others suggests that primary affective modes are *subjective but unconscious*; i.e., there is something it-is-like to be in them, but subjects (including humans) are often unaware *of* what-it-is-like. The real achievement of affective neuroscience, then, is the discovery and documentation of unconscious subjectivity in animals. What Panksepp calls "affective consciousness" is actually part of the animal's first-person orientation and engagement to the world of concern. This baseline form of mental life is *for* the animal and *from* its point of view, even if the animal does not know this or represent it.[25]

Obviously, the general idea of a primary affective mode is not new to psychology any more than it is to neuroscience. In some respects an affective mode is akin to the Freudian concepts of "drive" and "primary process." Within the psychoanalytic framework, these were conceived as irrational impulses at the heart of the psyche. Neuropsychoanalysts sometimes appeal to Panksepp's work as empirical vindication of Freudian metapsychology (Solms, 2003). For example, the ML-DA SEEKING system might be thought of as a *new pleasure principle,* a "reward system" for ego psychology. This is interesting but the similarity here is easily overstated. SEEKING does not aim at wish fulfilment, but at learning about and engaging the world. It is not opposed to any "reality principle." It is not something that is repressed. It is not deeply or essentially sexual. But in one key respect psychoanalysis is right where affective neuroscience is wrong: primary affective modes can be and often are unconscious.

Unconscious subjectivity is more closely related to ecological psychology than to traditional psychoanalysis. The first-person perspective is governed by learned (mapped) affordances specific to the animal's environment, body, and history. In turn, the relevant first-person valences are situated within phylogeny. To illustrate the contrast between the psychoanalytic unconscious and the unconscious subjectivity anchored in the mesolimbic system, return to the example from the end of Chapter 3. My friend's unconscious avoidance of his old neighborhood was not "neurotic" behavior, except in the thin sense that it was unreflectively driven by the neural systems of affect. It was not libidinal in any specifically Freudian sense. It is not particularly helpful or insightful to think of the stolen guitar as a phallic symbol. There was no inner struggle between illicit drives and superego rationality or morality. Sublimated psychic energy did not find cathexis or release in some disguised form. No. The whole thing is better seen as a case of the navigation of a learned layout, governed in part by affective valences of which the subject was unaware. My friend's cognitive map had a feeling of tone or texture, guiding his navigation (or: his unconscious emotions had a spatiotemporal structure). There was a negative value at a certain place in his "geopsychic" landscape, reflecting an emotional memory. All this was for the subject and part of his first-person mental life, but remained unconscious.[26]

Luckily, Panksepp's key concept of a primary affective mode can be extracted from the rest of affective neuroscience fairly cleanly, and repositioned within the larger evo-devo framework for baseline subjectivity introduced in this chapter. This is because Panksepp's idea was originally constructed within an evolutionary framework. Panksepp always emphasizes the shared evolutionary history of all mammals, the extensive homology in the mesolimbic system, and the comparable behavioral responses across these species.

3.2 Connecting OS Hypothesis and primary affective modes

Finally, it must be emphasized that the mechanisms of subjectivity don't operate in a vacuum. Map memory for navigation of the layout (OS Hypothesis) requires SEEKING mode as a basis for learning. Conversely, the SEEKING system is not just a state of arousal, unconnected to perception and memory. It is fundamentally linked to the navigation and memory systems – one cannot "seek" without a world to explore. This is so even in the open-ended and subintentional sense of the primary affective SEEKING system. And again, it turns out that the connections between these two subsystems of subjectivity are already studied, often under the rubric of "emotional decision making" or "expectation." The dorsal and ventral striatum appear to play an important role in linking spatiotemporal maps with the emotion systems. The hippocampal-striatal axis is emerging as a key and underappreciated pathway (see the end of Chapter 7). HP area CA2 has hitherto been a poorly understood component of the memory system, often left out of cognitive map models. Recall that the Parallel Map Theory locates the Bearing Map in the DG, the Sketch Map in CA1, and the integrated map in CA3. But there is also evidence of modulating connections between these areas and CA2. Emerging evidence suggests that activity in CA2 may feed back and forth with the striatum, which stands at the "top" of the SEEKING system, and is tightly associated with the lateral hypothalamus. Thus, the hippocampal-striatal axis may be a key linkage among the many subsystems of baseline subjectivity.

4 Toward a science of baseline subjectivity

There is already a wealth of knowledge about the neurobiological basis of the first-person perspective, and it is evident that a formidable science of subjectivity is on the way. As indicated early in this chapter, the core of for-me thought is an affectively valenced sense of directionality in an environment. There are numerous empirical models of neural mechanisms underlying this baseline form of subjectivity, including Jacobs' OS Hypothesis, Panksepp's primary affective modes, and Freeman's limbic loops. Of course, these models are still preliminary and incomplete, and will only form part of the emerging portrait of animal subjectivity.

With this much in place, I next turn to some of the philosophical implication and theoretical background of this work in neuroscience. Why think that these models of situated cognition help explain anything "subjective" or "first-person?" Why think that there is progress on the explanatory gap? What does all this mean for the debate about reductionism, eliminativism, and functionalism?

7
Putting the *Neuro* in Neurophenomenology

1 Introduction

How does the evolutionary developmental biology discussed in the previous chapter help explain, in Levine's phrase, how anything like a first-person perspective could arise in the world (Levine, 2006; 2001)? In this chapter I argue that the outline of an account is emerging, with the resources to explain not only how animals navigate the environment but also why the phenomenology of the first-person should be as it is, namely, an identification free perspective in a world of affective salience. The neurobiological image depicts mental life as a matter of "embodiment" – that is, as caused by the particulars of morphology and development. The evo-devo framework treats animals as developmental systems composed of homologues. Thus, to explain an embodied mind within this framework would be to show how and why the activity and functional interaction of homologues form the causal basis of the first-person perspective. The evo-devo research introduced in the last chapter makes significant progress toward such a causal account.

The science of subjectivity should strive to be *phenomenologically valid*. This means explaining key aspects of the phenomenology of the first-person, such as subjective temporality. Showing that the ecological function of subjectivity is navigation/coping does not, by itself, account for the distinctive phenomenology of the first-person. Why is subjectivity *for-me* in the way that it is? What is the specific difference between this form of navigation and others? Why doesn't a robot vacuum cleaner have a first-person perspective, since it too navigates an environment? Addressing these issues requires that the account do justice to our particular form of embodiment. In short, what is needed is a sort of *neurophenomenology* of the first-person perspective.

A key idea in what follows is that of *homology*. Biologists often explain a given character by citing its history and development, not its function, and this explanatory strategy has been dubbed *homology thinking* (Ereshefsky, 2012). Within the neurobiological image, homology thinking fills out the notion of embodied subjectivity, seeing it as a historically conditioned moment within a developmental system composed of homologues and their interactions. This vision of *modular embodiment* offers a form of neurophenomenology that is directly continuous with affective neuroscience and evolutionary developmental biology (not merely consistent with them). As an example of the explanatory potential of a theory of modular embodiment, I suggest that *temporality*, a key element in the Husserlian analysis of subjectivity, can be explained as the affectively valenced "slope" that results from the particular way animals like us navigate our world. This explanatory payoff stands in sharp contrast with the story offered by enactive neurophenomenology.

The first part of this chapter introduces a stringent criterion for any putative explanation of subjective phenomenology. I then move to introduce modular embodiment and the phenomenology of subjective temporality. In the end it should be plausible that the proposed criterion can be met for the case of temporality.

2 Jackson's Constraint: a priori physicalism for the sake of argument

How can we tell whether there is progress in explaining subjectivity? Part One of the book is largely dedicated to establishing that an empirical explanation is at least possible. But some criterion is needed. It would be best if there were a criterion that all parties could agree to, even skeptics. Fortunately, one of the leading skeptics about the prospects for explaining consciousness has provided just such a stringent measure. Frank Jackson (2003) has formulated a constraint that any scientific account of experience ought to strive to meet:[1]

> *Jackson's Constraint:* The account should allow us to see how the passage from the physical to the nature of experience might possibly be, somehow or other, a priori.

According to this criterion, the scientific explanation should indicate a path or "passage" between the objective and the subjective, such that an ideal knower could in principle know the content of subjective experience from purely empirical data plus the correct functional or conceptual analysis. The constraint does not require that any actually existing individual could make this passage.[2] Rather, it requires that the scientific account should provide insight into why the subjective character of experience should be as it is. So, Jackson holds that progress in explaining consciousness will occur only if we begin to make the relation between experience and the brain *intelligible*. Jackson's Constraint is the standard of *a priori physicalism*.[3]

Jackson provides an example of the kind of explanation he has in mind:

- Temperature in gases is that which does so & so (a priori premise about the concept of temperature);
- That which does so & so is mean molecular kinetic energy (empirical finding);
- Therefore, temperature in gases is mean molecular kinetic energy.[4]

The argument makes an identity claim, grounded in a functional definition, which provides a link to the mechanisms at the underlying level. Thus, using only the functional or conceptual first premise ("a priori"), one can make the passage from the empirical second premise to the informative identity claim in the conclusion. Thus, the example generates a schematic form for explanations that will satisfy an a priori physicalist. Meeting this constraint by filling in the scheme for some aspect of experience would constitute progress in the search for a science of consciousness.

Needless to say, a priori physicalists such as Jackson, Levine, and Chalmers do not expect this criterion to be met in the foreseeable future. And so far they are right about at least one primary aspect of consciousness. Phenomenal quality (e.g., the *redness* of red experience) has proven recalcitrant because the functional premise cannot be formulated. The redness of red has not been shown to have any function whatsoever. But subjectivity is a different case; we know what for-me-ness *does*, and an ecological function for the first-person perspective can be specified. Further, we also know quite a bit about *how we do it*. These are steps toward meeting Jackson's Constraint, which can then be used in the search for insight into why the first-person has the character is has.

2.1 Meeting Jackson's Constraint for the case of subjectivity

Here is the shape of the argument:

- Baseline subjectivity is identification free, affectively valenced spatiotemporal orientation in the environment of concern.
- In mammals, identification free, affectively valenced spatiotemporal orientation is explained by a set of dynamic limbic mechanisms.
- Therefore, baseline subjectivity in mammals is explained by a set of dynamic limbic mechanisms.

The argument consists in a functional analysis of the basic structure of the first-person point of view (the first premise), connected to a set of functional models in neurobiology (the second premise), issuing in the explanatory conclusion. I briefly comment on this schematic presentation before taking up the phenomenological questions below, in the longer sections on homology thinking and temporality.

The first premise was argued for in the earlier chapters of this book. According to Jackson's terminology, this is the "a priori" part of the argument. Though the first premise is surely not a logical truth, it counts as priori in two relevant respects, also illustrated by Jackson's example with temperature. First, it is a functional characterization that does not rely on any particular finding in neuroscience. Second, it is a conceptual characterization directly drawn from the analysis of the first-person perspective as an identification free form of reference. The empirical claim in the second premise was the subject of Chapter 6, except that now the identification free nature of animal navigation is explicitly stated. So far, the point is just that the first-person perspective *does something*. Its function is orientation, navigation, and learning. This serves to connect it with the neuroscience and fill in the argument scheme given in Jackson's Constraint. (Again, more will be needed to render this claim phenomenologically valid, and this will be the topic later in the chapter.)

Notice that the second premise is couched as an explanatory claim, not an identity claim as in Jackson's example, and this affects the conclusion. The argument does not purport to show that subjectivity is a set of dynamic limbic mechanisms; only that subjectivity is *explained* by them. The first-person perspective is an organism-level property, but its causal/historical explanation will devolve on particular component mechanisms of the organism, which do not themselves have the property.[5]

This subtle departure from the argument scheme for a priori physicalism will not block the claim that Jackson's Constraint can be met.

The upshot is this. In Jacksonian terms, we are beginning to see how the phenomenology of the first-person "might be, somehow or other, a priori." The second premise fills in part of the story about the "somehow" in the "somehow or other" phrase. Since the first-person point of view is, at base, an animal's affective and perspectival orientation in its environment, we can identify the neural mechanisms that enable and maintain this orientation. Discussion will indicate that dynamic models like those offered by Panksepp and Freeman *do* begin to render the relation between subjectivity and the brain intelligible, helping to clarify what it means to say that consciousness is always *for* some subject. At the baseline level, there is no further representational step to *get to* the first-person perspective (no "identifying judgment," see Chapter 2). For-me-ness falls out of affectively guided navigation directly. Subjectivity just is the first-person perspective of an animal like us. This perspective consists in spatiotemporally articulated affectivity. Our feelings themselves are perspectival, and this is what makes them for-me. The world of subjective experience is neither the external world itself nor a third-person domain of conceptual knowledge. It is the directly lived world of ongoing subjective concern. Baseline subjectivity is necessary but not sufficient for complete human experience in its full qualitative, symbolic, and socially mediated splendor. But in fact Jackson's Constraint is not so impossible to meet. We are beginning to explain how, in Levin's words, something like a subjective point of view can arise in the world.

3 Homology thinking

So, is the human hypothalamus like the rat hypothalamus? Pretty much. The basic anatomical differences in hypothalamic structures of male and female lab rats are also seen in humans. Nevertheless, human sexual and social behavior is vastly more complex. Your very large brain, especially your huge prefrontal cortex, means that your flexibility in navigating your social world and your capacity to control sexual behavior is far more rich and varied and involves much more learning than in the case of a rat. In large-brained mammals, the interaction between gene expression and learning-based changes in the brain becomes a dense thicket of complexity. – P.S. Churchland (2013, p.137).

3.1 Preliminary on reverse engineering and evolutionary psychology

A common strategy in cognitive science is known as "reverse engineering." Typically, the goal of reverse engineering is taken to be the invention of a model replicating an I/O function of some target system, such that the model reproduces system behavior: "The purpose is to deduce design decisions from end products with little or no additional knowledge about the procedures involved in the original production" ("Reverse Engineering," Wikipedia, 3/31/13). *How* the output of the model is produced is strictly a secondary question; what is important is the end product. Reverse engineering taken in this standard sense is really no different from plain old engineering, except that there is already in view an existing system that produces the desired output. This emphasis on products over processes makes the standard conception of reverse engineering rather ill-suited to neuropsychology, in which the aim to discover *how the brain does it*. But sometimes the phrase "reverse engineering" is used in an explanatory context wherein the goal is precisely to produce the behavior the way the target does. This is a very different task, one that obviously requires lots of knowledge about the original (cf. P.S. Churchland, 1994). To explain how the target system works is to be able to cut the system at the joints, to know the particulars about the underlying mechanisms.

Evolutionary psychology in at least one of its popular forms is characterized by the first and more familiar kind of reverse engineering. "Traits" are individuated by selected functions. These are identified externally in the sense that there is no attention to how the animal does it. Instead, the process most directly in view is natural selection, which sorts traits according to output function. So, the proper parts of organisms turn out to be the assortment of specialized output functions, or phenotypes. This picture is extremely powerful, and within philosophy of biology it is closely related to *population thinking*, the classic slogan coined by Mayr (1959). "Population thinking" captures the essence of the concept of relative fitness, and so it is one key to understanding evolution by natural selection. Population thinking appeals to the structure of a population to explain maintenance of a trait under selection. This makes it clear that no trait is intrinsically adaptive (i.e., intrinsically functional under selection). Instead, a given phenotype is more or less fit relative to a distribution in a particular population, living in a given selective environment. Both population structure and selective environment change significantly over time, so evolution by natural selection is a deeply contingent affair. But more to the immediate point, it is a deeply functional affair. The lesson of population thinking is that relative fitness is strictly a matter of differential reproduction in local contexts. Thus, population thinking is a way to employ reverse engineering (in the first and more standard sense) within biology, and the result is evolutionary psychology in its more familiar forms. The approach is "ahistorical" in one very specific way. The structure of the population, and hence relative fitness of any given trait, is a *current* fact. How the population and the trait got the way they are is irrelevant. Thus, mainstream evolutionary psychology can explain the distribution of traits as a function of population structure over time, which is of course "historical." But if the organisms were suddenly replaced by robots or zombies that performed the same output functions *with completely different neuropsychology*, this difference would be invisible to the evolutionary psychologist.

So it is the second meaning of "reverse engineering" that is relevant to evolutionary developmental biology and the explanation of embodied subjectivity. The evo-devo approach to reverse engineering has been characterized as *homology thinking* (Ereshefsky, 2012), a phrase meant as a contrast and complement to Mayr's notion of population thinking. The project of putting the neuro in neurophenomenology should be an exercise in the kind of reverse engineering that can be directly informed by the history of the bodies that perform subjectivity, that is, by the way the animals actually do it. This would be an "embodied" approach in best sense, and the concept of homology is essential to the project.

3.2 Homology, analogy, activity

170 years ago Richard Owen spoke of a homologue as *"The same organ in different animals under every variety of form and function"* (Owen, 1843, p.374). Antedating Darwin, this original definition makes homology independent of fitness and explicitly distinguishes homology from function. The contemporary literature on homologues continues Owen's tradition by treating them as *recurring structures* that can appear across levels of organization, from gene networks and epigenetic pathways to morphology and behavior.[6] Homology is a fundamentally comparative and historical notion with both developmental and phylogenetic elements. Recognition of homology at the psychological and phenomenological levels is the first step toward formulating the account of embodiment needed to naturalize subjectivity.

A basic distinction for homology thinking is between the homologous *character* and its *character state* (Brigandt, 2007).[7] One and the same character can exist in many different and qualitatively dissimilar character states across individuals and species. The identity of the homologue (its "sameness" across species) is determined historically. Two structures are homologous if they are descended from a single structure in a common ancestor. In this sense, homologues are the units of phenotypic transformation. The different character states of one and the same homologue can have very different adapted functions, while unrelated and non-homologous characters can have the same adapted function.[8]

Even from this quick introduction, it is readily apparent that there are interesting implications. For example, critics of evolutionary psychology (and its allied research in behavioral neuroscience) rightly object to oversimplified claims that, e.g., pair bonding in prairie voles is "just like" human monogamy. The distinction between a homologous character and its various states allows this objection to be met and points the way toward a better, more nuanced comparative psychology. It shows how it is that although the mechanisms of baseline subjectivity are homologous across vertebrate species, this does *not* entail that the experience of (e.g.) prairie voles and humans is "the same," qualitatively, developmentally, or even adapted-functionally. In fact homology thinking helps specify the *differences* between us because homologous characters form *lineages* in much the way that species do. Lineages are not designed for anything at all, and their function and character are not what individuates them. Yet relationships of lineage explain certain similarities and differences among the related organisms.

A simple example of a homologue is the mammalian forelimb. Different character states of the homologue include the arm of a human, the front flipper of a whale, and the wing of a bat (Brandon, 1999). In turn, the mammalian forelimb can be seen as the character state of an older and more general homologue, the vertebrate limb. Thus, one contemporary reformulation of Owen's definition of homology reads: "A homologue is the same part, such as the vertebrate limb, under every variety of form from shrew to elephant, and every variety of function from burrowing to flying." (Brigandt & Griffiths, 2007, p.4)

Homology thinking, then, explains observed character states in terms of their history and development, not via their selected function.[9] Explanation by selected function (population thinking) consists in showing how a particular feature helps accomplish some adaptive goal, and then inferring the relevant selective history. For example, wings are rigid because this feature has been selected for as a solution to the evolutionary problem of flight (Ereshefsky, 2012; Brandon, 1999). By treating flight as a design problem, reverse engineering shows that all wings must have this property if they are to perform their function, whether insect wings, bird wings, bat wings, or airplane wings. The shared functional properties of these wings are *analogous*. But because insect wings are descended from the gills of ancient crustaceans while bat wings are descended from mammalian forelimbs, insect wings and bat wings are *not* homologous (Carroll, 2005). This explains many observable differences between insect wings and bat wings (that the former come in pairs, for example), differences that are not comprehended by their analogy. Homology thinking seeks a stronger, deeper, and more directly causal explanation that appeals to the particular *historical paths* in the genesis of two kinds of wings. In the evo-devo framework, this historical explanation has two basic ingredients, trait polarity and developmental covariation (see previous chapter). To account for the wings of an insect, one must know what structure existed prior to the appearance of the wings, and how further events unfolded in the subsequent history of the homologue. Similarly, in order to explain the biological basis of the first-person perspective, we must study the historical path by which it arose.

Evolutionary psychologists will object to the idea that there is any explaining to be done beyond identifying the trait's contribution to fitness, or beyond the functional role played by the trait. When it comes to explaining flight, for example, they all have rigid wings. The rigidity is what explains the ability to fly, and it is that ability that contributes to relative fitness, so what's the problem? For the case of the first-person perspective, it might be argued, the neural mechanisms that enable navigation are what contribute to fitness, and the "embodied" historical differences are just plain *irrelevant*, mere implementation.[10] This objection misses the point – or rather, it illustrates the point. The particular way *we* do it involves subjectivity, and this is not a feature of the second-order (multiply realizable) functional analogy. The gap between first-person and third-person discourse arises precisely because subjectivity is *not* specified in the design solution. This little devil is in the historical details. The fact that the function can be realized without the subjectivity is what shows that there is more to explain; it is what shows that embodiment is not mere "implementation."[11] The real payoff for the thesis of our embodiment can only be found in the historical contingency of our situation and in the differences between different ways of being.[12]

Consider another example. Analogical reasoning (population thinking) does not, by itself, comprehend the differences between the external body shapes of tunas and dolphins (Pabst, 2000). Because of common environmental demands, the two species have very similar bodies, and population thinking certainly explains this as an evolutionary analogy. But there remains an "explanatory gap" between the adapted function and the actual animals that perform it. The same holds for the difference between the implementation of a function such as navigation in humans and other analogous mechanisms, or simply as modeled formally. For example, a robot vacuum cleaner could easily be designed to incorporate an "estimator" of its trajectory through the household.[13] But this functional analogy with temporal sequencing would not share the fine-grained structure of affectivity, anticipation, and memory that shapes our way of being in the world. This certainly doesn't mean that our special kind temporality is not embodied in a perfectly natural biological organism. It means only that the difference in the mechanisms is to be explained through historical understanding of the animal, its component homologues, and the developmental system in which subjectivity takes place.

Theodosius Dobzhansky's (1973) essay entitled "Nothing makes sense except in the light of evolution" is sometimes invoked by adaptationists who believe that population thinking and biological analogy are what make sense of *every* biological explanation. But the "light of evolution" isn't shed by the principle of natural selection alone; descent with modification helps make sense of things, too (Love, 2007, p.706). And this is what homology thinking brings to biological explanation. In an important elaboration of homology explanation, Alan C. Love extends the concept of homology beyond structures, such as skeletons, to include processes, such as the transcriptional action of a gene. Homology of process can be conceptualized in terms of *activity*. The activity of a given process is "how it works," which can be distinguished from that activity's *use*, whether this use be its causal role function (contribution to a capacity), or selected function (contribution to maintenance under selection). Thus, a process can be homologous in the required sense – it can be an activity that remains the same *under every variety of form and function* (Love, 2007, p.695). That is, one and the same activity plays different causal roles and different selective roles, depending on context. In this way, homology of function can be understood in terms of "what it is" or "what happens" rather than "what it is for" (see also Brigandt, 2007).

The idea of a process homology is essential to the project of explaining the first-person perspective from within an evo-devo framework. Subjectivity is what happens when animals like us navigate their environment. What the first-person perspective is *used for* is navigation, but this does not explain *what it is*, let alone how it works. Activity-function can provide an important conceptual connection between evolutionary development and behavioral neuroscience. As Love points out (pp.702–3), it grounds the argument for analyzing psychological and behavioral categories in terms of homology. It can now be seen that the proper parts of organisms are homologous characters, individuated by the organizational features of structure, activity, and causal role. In this way – and only in this way – we can identify neural *mechanisms* across phylogenetic distances.

3.3 Homology, evolvability, and developmental modules

Homologues are the proper parts of animals; they are the components of embodiment. How do they form an animal whole, and what role do they play in embodied subjectivity? The next relevant idea is that of a *developmental module* (Wagner, 1996). Homologues are not simply adult traits, the linear phenotypic "end products" or "outputs" of genetic algorithms. Homologues are constructed in each generation not only by genes and gene networks, but in the ontogenetic regulation of gene expression. Development is a process mediated by many stable and heritable (but of course varying) structures of physiology, ecology, and sociality (Wimsatt, 1997). Thus, a developmental module is a quasi-independent generative complex, a developmental subsystem that produces, enacts or constructs homologues over the lifespan of the organism. Homologue development is "modular" in the weak sense that changes can occur in one sequence without much coordinate change in others.[14] In this sense, homologues are "evolvable" because they possess some degrees of freedom relative to other parts of the developmental system. Mammalian forelimbs were able to vary and evolve relatively independently of the circulatory system, allowing them to respond to natural selection without killing the animal by derailing development. This is due to plasticity within each developmental subsystem and to the *weak linkage* (soft assembly, modular organization) between them (Gerhart & Kirschner, 2005). Circulatory development is constrained and scaffolded by skeletal and muscular forms, such that a change to the limb does not require simultaneous genetic changes in circulatory system. This "quasi-independence" of homologues is strictly a relative matter, and particular character states are progressively locked in during the course of development, creating an integrated functional whole with mutually dependent parts.[15] Many proponents of embodiment in philosophy of mind emphasize the holistic integration of the lived body, concluding from it that mechanistic explanation or "reduction" is impossible (e.g., Thompson, 2007). But using tools such as the genotype-phenotype map, multilevel developmental modules can be identified as bases for dynamic functional decomposition.

Developmental modules are highly conserved. Beneath the staggering diversity among vertebrate species, basic body and brain plans display remarkable invariance (Kirschner & Gerhart, 2005). The mammalian forelimb remains essentially the same character across its many modifications. Likewise, the vertebrate limbic system remains an essentially stable set of characters, despite the many states, combinations, and adapted functions in different species and individuals.[16] How similar are the systems, and how does this history of similarity and difference show up in the subjectivity of different animals? These are questions for neurophenomenology proper. Simple anthropomorphism is surely to be avoided, but there is ample evidence of deep homology (Strausfeld & Hirth, 2013; Kirschner & Gerhart, 2005). It appears that there is a kind of deep structure to the neurobiological image of embodied subjectivity. For creatures like us, there is an invariant core, a basic kind of thing that navigating the world is like. The core of the first-person perspective is a salience structure which forms a mode of *motivated learning*. I conclude the introduction of homology thinking with an example of this last thought.

3.4 Subjectivity in development, or, to the neocortex and back again

Why think that human subjectivity in its linguistic and sociocultural splendor is explicable (in part) as a specific character state of a deeper homology with non-linguistic species? Why believe that essential elements of human coping are shared with the birds and beasts? After all, we know that language and self-understanding in adults are mediated by the unique neocortex. Beyond a general allegiance to Darwin, then, how does evo-devo really help with the science of subjectivity? Begin by thinking of subjective coping (i.e., mapping with affective valence) as *motivated learning*. Syal & Finlay (2011) explore the mechanistic links between social motivation and language learning, and argue that homology thinking can be fruitful even for investigating a uniquely human trait such as language, underwritten by the uniquely human neocortex.

In essence, Syal & Finlay argue that language acquisition is a special case of vocal learning and sharing, facilitated by a three-way link between motivational systems, highly valued social relationships, and a flexible learning system in the neo-cortex (2011, p.426). In short, language acquisition is gated by the motivating properties inherent in social interaction (Kuhl, 2007). Basic mechanisms of language acquisition, then, may be homologous with those of vocal learning and vocal sharing in other social but non-linguistic species, including songbirds. In this story, the evolution of language was made possible by developmental adjustments to existing socio-motivational systems and vocal mechanisms. Here, language acquisition is partly a matter of learning to navigate a "space" of great concern, the rich domain of potential social activity.

Syal & Finlay's version of homology thinking, which they dub "developmental conservation," begins from the idea, described above, that a basic and evolvable brain plan had stabilized before the divergence of birds and mammals. They marshal evidence for motivational dependencies in social learning, connect this with current work in neuroanatomy (this is the homology part), and paint a picture of language acquisition as deeply situated, social, and affectively motivated. Highlighting the overlap between social behavior and motivation in the ML-DA system – especially but not only the ventral striatum/nucleus accumbens – they argue that vocal learning is driven by a *desire to engage* with valued individuals such as caregivers, not by simple stimulus/response pairings or the automatic unfolding of a genetic program (p.422). The idea is that developmentally delayed neocortical systems, like those that underlie language and reflection, are informed by the prior *subjective* proclivities of spatiotemporal orientation and affective engagement. Using Panksepp's (1998) terminology, children SEEK to acquire language as part of a developmental process, that of learning to navigate their social world (see Chapter 6). In turn, this motivated exploratory activity is a *causal* factor in the growth of the neocortex. The first-person perspective, then, is a developmental module (set of modules) that has come to play a key role for an individual animal whose ecological context is that of an "obligatorily social, gregarious, and often-altruistic species" (p.420).

For present purposes, Syal & Finlay's research illustrates several things. First, it illustrates the way homology thinking guides research in cognitive evolutionary developmental biology, as a complement to population thinking. Second, it points to the learning and motivational functions ("mapping" and "affective engagement" respectively) of the first-person perspective. Third, it offers a contrast to the common picture of the limbic system as the merely subpersonal slave of corticothalamic consciousness and cognition (see for example, Tononi, 2012).

Syal and Finlay's work demonstrates the value of the idea that baseline subjectivity can be shared across diverse ecological and behavioral contexts without being "the same" in an overly simplistic sense. Within developmental context, homologues can be regarded as the proper parts of animals, performing consistent and comparable "activities," even while each animal is treated as a distinctively integrated whole. Thus, the thesis of embodiment can take on a new interest and greater specificity. I now shift gears to discuss the embodied approach to mind more directly.

4 Modular Embodiment

The thesis of embodiment comes in many forms. Most proponents take themselves to be opposing something about mainstream computational cognitive science, but what this opposition amounts to varies considerably. A self-described "radical" version of the thesis is championed by Hutto (2012). He presents the embodied approach as a complete rejection of the representational theory of mind. In his view, thinking is nothing other than perception and action, which can only be fully explained in non-representational ways.[17] By contrast, I suggest that embodiment should be understood as complementary to the representational theory, not as opposed to it. The account of embodiment should play the role of a *theory of ground* for mental representations (Von Eckhardt, 1993). Embodiment is how thoughts acquire content for-me. The task of a theory of embodiment, then, is to complete the philosophy of mind by explaining the subjectivity of mental representation.[18]

This conception of embodiment – as a thesis about the ground of subjectivity – can be understood in relation to two other, more familiar approaches to embodiment, a functionalist version and a phenomenological version. *Embodied functionalism* is ably represented by Andy Clark (2012, 2001, and elsewhere). For Clark, the thesis of our embodiment is cashed out as *minimal cognitivism*. Mental states are still analyzed exclusively in terms of function, but realizers are distributed across brain, body and world, calling on fewer and less robust neural computations than standard cognitive models suggest. There is much that is right about embodied functionalism of this kind, but it fails to make contact with subjectivity or lived experience.[19] Clark's approach renders embodiment a footnote to mainstream philosophy of mind, not a conceptually significant complement.

By contrast, *embodied phenomenology* is a general approach propounded by Evan Thompson, Shaun Gallagher, Hubert Dreyfus, and others, and often called *enactivism*.[20] Here embodiment is taken in the tradition of Merleau-Ponty, cashed out via the phenomenological concept of "leib." Leib is *body subject* or lived body, and its contrast is "korper," the body taken as object of biological science. The problem of the explanatory gap between first & third person discourse is then construed as the problem of relating leib to korper. In embodied phenomenology, this gap is to be bridged formally and descriptively by topic-neutral DST redescriptions, a project that has been dubbed *Neurophenomenology*. The formal description of the relation between leib and korper is then to be placed within a more encompassing enactivism that attempts to provide a general ontological account of the emergence of leib. Through self-organization or *autopoiesis*, it is argued, mere material bodies become selves that have the sorts of holistic unity and subjectivity found in the phenomenology of the lived body. Thus, enactivism attempts an ontological passage from bodies themselves (korper) into leib.[21] The result is that, epistemically speaking, there *is no going back* from lived embodiment to mere korper, in the sense that an enacted unity (leib) can no longer be comprehended as a mechanical object (korper). Operating as a sort of principle of neovitalism, autopoiesis transforms bodies into a new dimension of being that cannot be comprehended by third-person discourse. This renders the relation between leib and korper much *more* mysterious, not less. Far from rendering subjectivity in the neurobiological image, enactivism renders neurobiology in the image of subjectivity. Consciousness, it is argued, is the condition for the existence of the "objective" world studied by science, and cannot be a natural part of that world. Ultimately there is a commitment to a transcendental philosophy of consciousness that stands opposed to a naturalistic worldview (I take up this thread again in the postscript on neurophilosophy, naturalism, and subjectivity). So, unlike embodied functionalism à la Clark, embodied phenomenology is conceptually different from mainstream analytic philosophy of mind and science. Further, its emphasis on subjectivity and the lived body is well placed. But it is neither explanatory nor naturalistic.

An approach to embodiment of the kind that I require – that is, a theory of the ground of subjective representation that meets Jackson's Constraint – must lie between embodied functionalism and embodied phenomenology. In short, the thesis of embodiment should begin to make the passage from korper to leib intelligible in naturalistic terms. A modular version of the thesis of embodiment can help accomplish this task. The idea is to draw on the evo-devo framework, in particular the concept of homology, to render an image of the lived body that can admit of explanatory analysis by causal mechanism. In this approach, the embodied "implementation" of functions will be given a strong explanatory role. The project of identifying the component parts of the organism (homology thinking) proceeds in parallel with the project of finding the functional organization (population thinking). So the functional analysis must always be understood relative to a particular embodied context, wherein "embodied" means *historically particular*. Dynamic systems theory (DST) can help here. Proponents of embodiment often appeal to DST as an alternative to "computation" or representation in cognitive science. This move is contentious, so I pause to introduce DST and connect it with the present theme.

4.1 Dynamic systems, functional decomposition, predictive coding

A dynamic system is any system that maintains coordination of a three-place relation between pairs of quantities in a range of contexts. Quantity A is coordinated with quantity B just in case B covaries with A in C (Ishmael, 2007).[22] Coordination is a "dynamic" process in the sense that the variables are mutually dependent; the value of each at a given time depends on the values of the others. This recursive structure means that changes in one variable can create non-linear or high-order changes in the system's overall behavior, not just changes in the next variable. Common examples of dynamic systems include the solar system, the digital computer, and the Watt steam governor (Van Gelder, 1993). For this discussion, it will also be helpful to draw a distinction between formal *models* and concrete *mechanisms*. System dynamics can be modeled by a set of differential equations, defining the evolution of parametric values over time. Strictly speaking, this abstract mathematical *dynamic model* is topic-neutral. By itself, it does not refer to, represent, or explain anything whatever. The model becomes explanatory only when it is extended *as* a model of some real world domain, some concrete *dynamic mechanism* (set of mechanisms). When DST is used in an explanatory context, the goal is to find and model the actual causal structure of the concrete system, cutting it at the joints. To do this is to engage in reverse engineering in the second, desirable sense described earlier in this chapter. For example, the geocentric and the heliocentric models of the solar system provide output functions that are equally predictive for a terrestrial observer. But they are not equivalent as explanations. Only the latter can be accurately extended as a model of the causal structure of the solar system.[23]

The abstract dynamic model can be represented as an n-dimensional *state space*, where n = the number of parameters in the model. Together, the momentary parametric values determine a point within the state space, and their differential evolution over time describes a trajectory through the space. In this way the behavior of the system can be plotted ("computed") within the space of possibilities. This formalism, however, is highly general. Since the term "computable" simply means *mathematically specifiable*, every model in neuroscience (or in any science) can be labeled "computational" and plotted onto a state-space. This does not serve to distinguish among varieties of concrete dynamic mechanisms, and this causes significant confusion. Gravitational systems, ecosystems, neural systems, hydraulic systems, digital computers, and communication networks can all be described mathematically in exactly this way. By this standard, even Hutto's (2012) anti-representational approach to embodiment would count as a computational theory. But in cognitive science the term "computational" is meant to be a claim about the concrete target system, about how it works, not about the formal model describing and predicting system behavior. Therefore the designation "computational" is sometimes reserved for *function evaluators*; i.e. those concrete mechanisms that implement formal procedures to decide whether a given information string belongs to a designated representational set (Mitra & Bokil, 2008, pp.37–8.). Computing can then be distinguished from *control theory*, which is a more general branch of engineering concerned with the design and functional implementation of dynamic systems. On this basis, non-computational control systems are often called "controllers" "regulators" or "governors" in order to distinguish them from function evaluators (Van Gelder, 1998).

The immediate point is that formal dynamic models (which are computational in the wide sense of being mathematically specifiable) can help pick out internal structure in the concrete target system *whether or not* the target is a "function evaluator" (computer). DST, then, brings the engineering virtues of functional decomposition that are the strength of traditional computational models, but without assuming well-defined information strings in the target mechanism. Exactly which set of equations accurately models the *causal* structure of the real world system is an empirical question to be tested by staging manipulations. But functional decomposition per se need not entail that the underlying dynamic mechanisms are "computers." Put the other way around, if the actual mechanisms are not computational, this does *not* mean that they are not functional or that their functional interactions cannot be modeled mathematically.

In computational neuroscience, one promising variant of dynamic modeling has come to be known as *predictive coding* (Howhy, 2013,

2012; Clark, 2012). The predictive coding framework incorporates elements of both the computational and non-computational control systems just distinguished. One general insight from DST is that the mechanism may be "circular" in the sense that it need not begin with "input" passively received and end with "output" or behavior (cf. Dreyfus, 2012). That is, the system's evolution over time results from the mutual feedback between state-dependent variables, not from the one-way transformation of input to output. The predictive coding approach incorporates this insight elegantly, by conceiving of "input" as *prediction error*. Sensory information is actively sought by the animal as part of a larger exploratory learning process already underway. Here, cognition becomes the dynamic coordination between expected values and feedback. Thus the predictive coding approach can be understood as a kind of "minimal cognitivism" that incorporates elements of both information processing and control theory (Clark, 2012). On one hand, the mechanism is computational. It functions by deciding whether a given information string (sensory feedback, error signal) belongs to a designated representational set (the expected values). But the given information string needn't be a well-defined and static data structure. It can take the form of continuous feedback which is almost always an "error" – i.e., poorly or incompletely defined – but this not render it useless. Since the designated representational set can be quite minimal and open-ended, specifying only a gist or a dimension of variation, the "decision" about the relation of input to expectation is a matter of real time pattern matching and adjustment rather than syntactic transformation. So, predictive coding is a great resource for Clark-variety embodied functionalism.

Among other things, the success of the predictive coding framework illustrates that dynamic systems theory is gaining ground as a general form of explanation, making inroads into the mainstream. Spivey (2007) argues there is an ongoing metatheoretical shift in cognitive science as a whole toward a *continuity framework*. This adjustment is required, Spivey argues, not only by developments in neuroscience but also by a converging mass of research from psycholinguistics, network modeling, and ecological psychology. Spivey predicts that cognitive science will adopt a new, more minimal concept of neural representation as *internal temporal dynamics*. I adopt this dynamic meaning of "representation" here, noting that it is a thoroughly functional notion. DST is an absolutely standard tool in the "systems" tradition of biological explanation.[24] Its growing influence on cognitive science is a theoretical reflection of the neurobiological image of mind and person.

4.2 Modular embodiment between function and enaction

Following Maturana &Varela (1980), Thompson (2007) and other enactivists reject functionalism on broadly holistic and emergentist grounds. Paralleling the phenomenological view of the holism of conscious content (with which I agree, see Chapter 8 on the NCC), Thompson suggests that the brain, too, must also be understood holistically, as a non-decomposable system (Thompson, 2007, pp.422–3). Since the nervous system constitutes a self-organizing (autopoietic) dynamic system that maintains its own structure and boundaries, Thompson holds that there is no value in talk of "pre-existing" components discoverable through experimental manipulation or reverse engineering. For example, he opposes localization studies in neuroscience because they do violence to the prior holism of the embodied creature. Any alteration or manipulation of the system creates a new unified whole, from which inferences about the original system cannot be made. Thompson quotes Maturana & Varela here:

> There is intrinsically no possibility of operational localization in the nervous system in the sense that no part of it can be deemed responsible for its operation as a closed network, or for the properties which an observer can detect in its operation as a unity. (Maturana & Varela, 1980, p.129)

But homology thinking shows the way past this problem. The investigation of the causally relevant components need not presuppose that they "pre-exist" the system in a way that is incompatible with self-organization.[25] Homologues are constructed in each particular animal's lifetime, in self-organizing developmental modules. But they also have a lineage of their own, one that does indeed "pre-exist" the particular organism, and that facilitates, constrains, and partly *explains* the "emergence" of the organized whole. Homologues carry the weight of their own history and their own dynamics into the embodied situation. The action and interaction of homologues is indeed responsible for the "properties which an observer can detect in its operation as a unity."[26]

The basic characterization of the nervous system (and the organism as a whole) as an actively self-organizing system is both true and interesting, and it is certainly important to acknowledge the difficulties of inferring local function in any system with the degree of connectivity of the brain. But the autopoietic unity of the nervous system does not constitute a kind of monad without component parts. Although it is true that temporal dynamics have often been neglected in mechanistic philosophy of biology, there is no need to choose between mechanistic (bottom up) and complexity theoretic (top down) approaches (Bechtel & Abrahamsen, 2011). Holism and mechanism are not any more incompatible than are nature and nurture. A lesson of evo-devo is that the two are not only compatible but in fact individually necessary and mutually supporting aspects of the total account.

The true meaning of the thesis of embodiment is that Leib and korper are the same. The difficult implication is that subjectivity really *can* be explained with the tools of objective biology. While qualia remain mysterious, there is a way forward for the explanation of the first-person perspective. Because the component parts of the brain have a history, we can cut the embodied mind at the joints. To explain an embodied mind, then, is to show how the activity and interactivity of homologues form the causal basis of the first-person perspective. To do this would be to provide a proper *neurophenomenology*. In the remainder of this chapter I indicate, in very broad outline, how to make a start on this task. One essential element in the first-person perspective, analyzed and elucidated in the phenomenological tradition, is *temporality*. So temporality is one key explanandum for the science of subjectivity.

5 Temporality considered as the affective dynamics of the first-person

A brief detour into the phenomenology of temporality will establish what needs to be explained. Subjectivity doesn't just take place in time or at a time. It is a temporal articulation *of* intentional content. Husserl called this dimension of experience *inner time consciousness* and analyzed it as a continuous structure of *protention* and *retention* within the first person perspective. His analyses appeared in his working notes, and remained exploratory and incomplete. In addition, he returned to the topic of temporality later in life, approaching it through the lens of "genetic phenomenology" (Depraz, 2008; Steinbock, 1996).[27] This is not the place for careful exegesis of Husserl's many texts and their final verdict on temporality. To simplify matters, I work directly from a contemporary, canonical, and unusually compact presentation of Husserl's analysis, written by Gallagher & Zahavi (2012) specifically for a cognitive science audience:

> Husserl employs three technical terms to describe this temporal structure of consciousness. There is (1) a "primal impression" narrowly directed toward the strictly circumscribed now-slice of the object. The

primal impression never appears in isolation and is an abstract component that by itself cannot provide us with a perception of a temporal object. The primal impression is accompanied by (2) a "retention," or retentional aspect, which provides us with a consciousness of the just-elapsed slice of the object thereby furnishing the primal impression with a past-directed temporal context, and by (3) a "protention," or protentional aspect, which in a more-or-less indefinite way intends the slice of the object about to occur thereby providing a future-oriented temporal context for the primal impression.[28]

The protention-retention structure of subjectivity is a "direct" intuitive grasp of the about-to-happen and the just-past; it is the meaningful flow of the stream of consciousness itself. What is "direct" about it? It is for-me in the identification free sense discussed in Chapter 2. That is, the question "whose" temporality this is does not arise and cannot be answered through the flow per se. I return to this interpretation in a moment.

The first thing about Husserl's three-fold analysis that bears comment is that the term "primal impression" is remarkably unfortunate because it is neither *primary* nor an *impression*. Gallagher & Zahavi clarify that, in consciousness, there is no knife-edge "now" moment, no experienced yet durationless impression that has a primary presence and that is afterwards compared to an expectation and a memory. "Now" consists in a gestalt of protention-retention, within which ongoing sensory feedback is situated. It would be just as accurate to characterize protention and retention as "primary" relative to sensory input at a given moment. The "primal impression" should be understood as a minimal line of tension in the structure of differentiation between future and past.[29]

Protention and retention are "peculiar forms" of intentionality because the subjective sense of temporal flow does not take its own, separate object (it is not the consciousness *of* time), but instead is part of the "microstructure" of the first-person itself, the way of apprehending one's temporal relation to objects and events (Gallagher & Zahavi, 2012, pp.86–88). This microstructure is variously characterized as a temporal horizon, gestalt, periphery, and context.[30] So temporality is part and parcel of the stream of consciousness, or what it means to have a first-person perspective at all. Thus, Husserl's analysis contains a fundamental distinction between temporality per se and the particular contents of experience. Temporality is to be understood as an *invariant dimension* of the subjective mode of presentation, specifying "now" in relation to the horizon of protention-retention. Thompson (2007) writes:

> We can distinguish within experience between what changes or varies and what remains invariant. The contents of the present moment – the particular things of which we are aware – arise and perish. But the present moment as a structure of awareness does not change or vary. No matter what we experience, it is always there, or rather always here. It is not a changing content of experience, but an unchanging structure of experience, the three-fold structure of primal impression-retention-protention. (Thompson, 2007, p.326)

Notice, too, that this "standing" structure *does something*. It orders particular contents within a larger framework, specifying a subject position. On this reading, "now" is not a singular term, naming something in experience. It is a function determined by local context.[31]

So, I take it that Husserl's working notes on temporality offer an analysis of the first-person perspective that is quite congenial to one pursued throughout this book. Protention-retention is the temporal aspect of the way animals like us are directly oriented to and engaged with the world, wherein baseline subjectivity consists. Neither "the subject" nor "the present" make an appearance *in* experience. In fact they are absent (Derrida, 1973). Rather, the identification free structure of for-me subjectivity partly consists in the "here-and-now of things," which is specified without being explicitly represented.[32] Contemporary Husserl scholars provide a similar interpretation of temporality. Zahavi (2005a, 2003b) argues that the object of Husserl's analysis is *prereflective self-consciousness* (PRSC). According to Zahavi, subjectivity consists in a kind of intransitive (non-objectifying) self-awareness. Although I have argued in chapter four that subjectivity is not best understood as "awareness," Zahavi's phenomenological analysis is otherwise very similar to my view of subjectivity.[33] We both hold that, in the first instance, subjectivity takes place prereflectively yet also in a way that remains for-me. For my purposes, "prereflective self-awareness" is baseline subjectivity, i.e., it is what happens when we live through events first-hand.[34]

If the science of subjectivity is to be phenomenologically valid, it will have to include an account of temporality that flows from the neurobiological image. One might think of such an account as an exercise in *neurophenomenology*. But there is already a literature that travels under this name. I must introduce it now.

5.1 Neurophenomenology®

The term "neurophenomenology" was coined by Francisco Varela (1996), and it is closely associated with the enactivism he propounded with Evan Thompson and Eleanor Rosch (1991).[35] Thompson's *Mind and Life* (2007) is the most complete expression of neurophenomenology to date.[36] In this literature, the stated aim of neurophenomenology is to "weave together" phenomenology and neurobiology and to "bridge" the gap between first-person and third-person discourse (Gallagher & Zahavi, 2012 [2008]; Thompson, 2007). But broadly speaking, this is *everyone's* aim who works on the science of consciousness. We all want to know how neurobiology and experience can be understood in a common framework (e.g., Metzinger, 2003). This open sense of the term seems best to me, referring to the general goal of finding a unified framework for consciousness and neuroscience. But within the enactivist literature it carries a more specific meaning. I designate this proprietary sense *Neurophenomenology®*.

First, the "phenomenology" in Neurophenomenology® refers specifically to the Husserlian philosophical tradition. Outside Husserl scholarship, of course, the term "phenomenology" has a more generic sense, simply referring to the way experience is. But here the aim is to connect Husserlian ideas and analyses with neurobiology. Second, Neurophenomenology® aims only to *bridge* the gap between consciousness and brain, not to close it. The spheres of phenomenology and neurobiology are to be brought into "productive relation" with one another and rendered coherent within the overarching ontology of enactivism. The goal, then, is not the goal I pursue here, to render subjectivity in any neurobiological image. Third, Neurophenomenology® is meant as a distinctive method for conducting empirical research, a phenomenological brand of experimental philosophy that recalls introspectionism.[37] The working hypothesis of Neurophenomenology® is that the dynamics of experience can be tracked both phenomenologically and experimentally (Thompson, 2007, p.366; Varela, 1996). Here, the idea is to use disciplined first-person methods to uncover phenomena (i.e., gather data) that can guide and constrain neuroscience. Practitioners of Neurophenomenology® have attempted to make a start on the project by looking for correlations between Husserl's brilliant and meticulous notes on inner time consciousness, the introspective reports of trained subjects, and EEG recordings.[38] In the next section I focus on the second element just identified, the idea of integrating or "weaving together" the phenomenology and neuroscience within a larger philosophy.[39]

5.2 Neurophenomenology® and the DST model of temporality

In what does weaving together these two domains consist? Basically, it consists in placing both within a single formalism, that of dynamic systems theory. Scientists routinely provide mathematical models of neural activity. The thought is that these same mathematical tools can be applied to the protention-retention structure of temporality, and thereby mediate between phenomenology and neuroscience. In effect, the dynamic model is regarded as a *topic-neutral description* that can be applied to both first- and third-person data. The idea of topic-neutral re-description was originally championed by J.J.C. Smart (1959) as a way forward for the central state identity theory, to investigate the empirical hypothesis that consciousness is a brain process. Topic-neutral descriptions are necessary because subjective reports cannot be *directly* translated into the empirical language of neuroscience. There is nothing in the experience of pain or of redness, nor in its phenomenological description, that refers to any neural state. Even if first-person and third-person descriptions really do pick out one and the same event in the world as identity theory postulates, they do so in different ways by reference to different properties of that event. They are different *modes of presentation*, as the saying goes.[40] As a strategy to address this problem, Smart suggested that conscious states might be rendered topic-neutrally, to allow for rapprochement with physiology. This is a useful way to think of the project of Neurophenomenology®. DST models are topic-neutral in the sense that they are just sets of differential equations that may or may not map onto *any* real world system. The empirical question then becomes, does the model extend to the target system?[41]

The strategy Smart suggested did not salvage the central state identity theory as a philosophy of consciousness. The tricky thing about topic-neutral descriptions is that they have to be *adequate* to the phenomenon. They have to pick out and describe subjective states in a way that makes no mention of qualitative properties such as the "look" or "feel" of the experience. Any such neutral language will inevitably abstract away from the conscious content. The problem, then, is that redescriptions only end up changing the topic and in this sense they are inadequate to the phenomenon. Notoriously, this is why consciousness is the really difficult part of the mind/body problem (Chalmers, 1996; Nagel, 1974). But it is widely agreed that for the *non*-conscious properties of mental states, topic-neutral descriptions can be had, in the form of functional characterizations. Content-bearing states can be picked out by what they do, and this makes it possible to identify them with whatever neural processes that do *that*.[42] So, the really desirable thing about functional analysis in cognitive neuroscience is that it provides a relatively clear, topic-neutral way to connect the mental phenomenon to the physiological mechanism (cf. Metzinger, 2003). But again, since qualia per se don't seem to have a function, they resist this way of developing the topic-neutral strategy, too.

Neurophenomenology® eschews functional decomposition, confining itself to mathematical redescriptions (dynamic models as discussed in part 3.i.a above). According to Neurophenomenology®, the now-phase of consciousness can be formally redescribed as a dynamic vector containing the "trace" of the just-past as well as a heading or direction in the state-space (Gallagher & Varela 2003, p.123). The standing structure of inner time consciousness is said to correspond to a formal feature of the system, i.e. to the "geometry" of phase transition.[43] To the extent that this story is defensible, it can only apply to the abstract mathematics of the redescription. This is fine as far as it goes, but it doesn't go far enough. Neurophenomenology® looks only for formal parallels between experience and neural dynamics. Thus, the holistic/emergentist commitment that drives neurophenomenology® has an ironic result: the neural system ends up getting treated *strictly* as an I/O device! But this is the very conception of reverse engineering that is so abhorrent to enactivism! The lived body becomes a black box out of which the temporal dynamics of subjectivity spring, wholly formed.[44] For neurophenomenology® the aim is *only* to use only the descriptive, abstract formalism of DST to "find" isomorphism (parallelism) between neural and phenomenological structures.

One deep problem with this sort of parallelism is that since the temporal dynamics of any process x can be formally modeled using DST, there will always be *some* formal isomorphism between subjective temporality and process x. This is because for every two processes, there is some mapping or other between them. The DST formalism, then, cannot distinguish between those processes that embody subjectivity per se and those that are just plain old temporal processes. For example, there will be a topic-neutral DST description of the tides at Liverpool, which can be shown to be isomorphic with the structure of protention-retention found in the phenomenology of the first-person. This should not be counted as tidal-phenomenology. Putting the neuro in neurophenomenology requires an *explanatory* role for neurobiology.[45]

And for the case of temporality, there is a way to get to the properly explanatory account. As we have seen, the structure of protention-retention that Husserl identified *does* have a function. So here is an aspect of the phenomenology of the first-person where dynamic models can make significant progress in providing topic-neutral *functional* descriptions. The first-person perspective keeps me spatially, temporally, and affectively coordinated with the world. Providing abstract models of this sort of coordination is just what DST can do, and extending them to causal/functional systems in the brain is precisely the business of neurobiology.

6 Putting the *neuro* in neurophenomenology

All the pieces are finally in place. What remains is to simply point to the locus of an explanation. Needless to say, what follows is highly speculative and subject to revision depending on the facts. The work of this book has been to transform the problem of subjectivity into an empirical problem, to identify theoretical tools for this problem (evo-devo framework, homology thinking, modular embodiment), and to indicate the probable domain in which to conduct research (the mesolimbic system). So, whether the following conjectures about the neurophenomenology of temporality are correct is an empirical question. But if we are on the right track then something like the following may well be true. In fact, if navigation is explained in something like the way suggested, then the protention-retention structure of temporality is a prediction of the general model.

Roughly, temporality is a matter of affective "slope" in animal navigation. As affective modes (such as the SEEKING system) predictively code and then update their values in coordination with the bearing map,[46] this generates a *sense of where you are* in affective terms; where you are heading and where you are coming from. These navigational bearings, then, *should* be first-person-temporal in roughly the sense described by Husserl. Because the bearing map is a dynamically functional coordinating mechanism, we can begin to see why this process should be identification free and for-me in the ways required for a phenomenologically valid science of subjectivity. The first-person perspective is a "position" maintained through coordination of spatiotemporal salience and affective process. In turn, this is explained by the predictive coding of navigational affects. Subjective temporality, then, is caused by a cognitively minimal anticipation and ongoing recalibration of the affective salience of sensory feedback.

Think of temporal slope as the *direction of change* in affective value in spatiotemporal heading. In this way it can be seen that affective salience is not a "+/- inference" that is added to spatiotemporal representation, on top of a subjectively neutral or purely "cognitive" map. Rather, subjective feeling itself is spatiotemporally structured. The ecological layout "directly" interacts with affect and vice versa, probably by way of the connectivity between hippocampus and ventral striatum. This means that protention and retention may be causally explained by the interaction of limbic modules functioning within the context of an embodied animal navigating its world of concern.

First, notice that the intentional content of protention does not directly concern external states of affairs, but rather their affective salience for-me. Will I be interested? Will anything change? Will I learn something? Will things get better? These "expectations" have traditionally been assumed to be represented in an internal world-model which is compared with the sensory feedback. It is this excessively "cognitivist" or "propositional" approach that Freeman rejected (see Chapter 6). But if there is cognitively minimal predictive coding for the expected *valence* of experience, rather than for specific representational content, this can explain the phenomenology of protention much more directly. The consciousness of the about-to-happen is quite general – boredom, excitement, curiosity, fulfillment, – regardless of what "it" is that is coming. This range of protention happens in all contexts, whether exploring the backside of an object, the other side of town, the course of a narrative, the structure of a melody. Similarly, retention is not so much a specific memory of some neutral fact discovered, but a persisting feeling about what happened. Am I on course? Steering clear? Upwind or downwind? Nearer? Further? A country mile? Temporality can be thought of as a simple (but subtle) form of metacognition. It is an index of *how things are going*.

"Gotcha!" says the much-maligned and underappreciated evolutionary psychologist. "This capacity for tracking how things are going is a great candidate for an adaptation. Surely it is easy to see that it would have been selected for!" Just so, one might reply, but that just does not tell us what we want to know. The desired explanandum is not the capacity but the phenomenology. Why does the exercise of the capacity happen *like that*? Why does it assume the form of subjective temporality, rather than some other microstructure? It is clear that there can be many different ways of tracking spatiotemporal position, many analogues for this capacity. If the selected function is navigation, why should this be subjective for us in the way that is? To answer these questions, an investigation of descent with modification is required. For, as long as the output function does the job selection per se will not bother over the differences. We also know that the navigating system is composed, at the modular level, of activities (homologies of function) that have been conserved and re-purposed progressively, recruited in slightly different ways in each species, developed in slightly different ways in each individual, producing a unique character state each time. But there is a deep structure, a baseline character, underlying the diversity of vertebrate, species typical, and individual character states. These character states interact with selection, of course. Navigation is not re-engineered de novo by each animal. Instead, the history of navigation is brought forward.

Our ancestor navigated by means of the olfactory bulb, and this slowly evolved into an affectively articulated limbic system. Key structural elements in the developmental modules composing this system include the ventral striatum, hypothalamic nuclei, entorhinal cortex, and hippocampus. Functionally, these modular components can be labelled the, e.g., the *bearing map* and the *primary affective modes* including SEEKING. These modules have characteristic activities (what they are, what happens), interacting dynamically to achieve navigation and drive learning, ultimately helping the animal cope with its situation. Through a detailed investigation of these characteristic activities and the predictive coding mechanisms that they constitute, we can begin to see how the microstructure of temporality is predictable.

An example. For subjective salience, it matters what the predictive coding computer *is*, not just what system function it performs, or what information it processes. In Chapter 6 I suggested that the bearing map is "direct" and "situated" in a way that sheds light on the first-person perspective.[47] But it is important that such predictive coding mechanisms themselves consist of component modules with their own characteristic activities. Kent Berridge (2012) argues for an *incentive salience* account of mesolimbic function in which predictive coding functionality is synthesized with equally important neurochemical *state factors* in the system that underlies motivation, learning, and reward. In the incentive salience model, motivation can come apart from learning because of biochemistry, such that "salience" is not only a matter of the associative prediction of future reward value (p.1124). The effects of state factors on the animal is a characteristic activity of the MLDA itself – a process homology – that is not captured at the abstract disembodied level of the predictive coding function. Berridge argues that incentive salience, mediated by neurobiological state factors rather than pure associative learning, is key to understanding fluctuations in motivation, desire and subjective decision-making (see also O'Doherty, 2012).

Because it explicitly connects affect and motivation with prediction, expectation, and cognition, Berridge's incentive salience approach is one promising avenue for future empirical research on the first-person perspective. Chapter 6 outlined several other important avenues for research within the framework of cognitive evo-devo. As far as brain structures, I suspect that a productive locus for neurophenomenological research will be the hippocampal-striatal axis. The ventral striatum may be a key fulcrum connecting the systems of affect and habit with the mapping and memory subsystem in hippocampus. Pennartz, et al (2011) argue that VS may be the key to generation of goal directed action, and they attempt to reconceptualize this pathway in a way that goes beyond a classical reinforcement learning architecture and challenges the traditional dichotomy of episodic (hippocampal) and procedural (dorsal striatal) systems (p.549).

7 Resumé: from a priori physicalism to the neurobiological image

In the course of the last two chapters I have indicated, broadly, how something like a first-person perspective might have arisen in the world. To do this required more than an account of how animals navigate and how this capability might have evolved. It also required an indication of why baseline subjectivity should take the form of an identification free articulation of spatiotemporal and affective salience. Pursuit of this objective took the discussion, in this chapter, through the thickets of philosophy of biology, embodiment, dynamic systems theory, and phenomenology. In the end the story falls somewhat short of the criterion set by Jackson (2003) and other proponents of a priori physicalism. On offer is not exactly a modal *derivation* of subjectivity from purely functional premises plus empirical findings. Nevertheless, it will take a determined anti-naturalist mind not to see that what we have here are resources that render the first-person perspective more *intelligible* and less completely mysterious. And this, I take it, is the most charitable interpretation of the point of Jackson's Constraint in the first place.

All of the material in the preceding two chapters builds on the analysis of subjectivity put forward in Part One. In particular, it assumes that subjectivity is, in the first instance, unconscious. Recalling the example and discussion at the end of Chapter 3, my friend navigated his way through the affectively salient world of his neighborhood, avoiding the site of negative value, without being aware of doing so. Temporality, too, can and does take place beneath the level of awareness. For subjects like us, there is something it-is-like to orient and navigate in space and time, whether we know this or not. Temporality is truly a matter of the "auto-affection" of subjectivity, and in this sense the portrait of animal subjectivity sketched here is a contribution to the literature on genetic phenomenology (Marabou, 2012). This portrait is not a well-defined theoretical entity, formally specified a priori, from which the explanation of subjectivity might be modally derived. It is an *image*, a thumbnail sketch that guides theory and research. I return to the neurobiological image in the postscript, stepping back to talk about naturalism and neurophilosophy more broadly. First I turn, in Chapter 8, to the topic of consciousness proper and its relation to neural process.

8
Neural Correlates of Consciousness Reconsidered

It is widely accepted among philosophers that neuroscientists are conducting a search for the *neural correlates of consciousness*, or NCC. David Chalmers (2000) conceptualized this research program as the attempt to correlate the contents of conscious experience with the contents of representations in specific neural populations. This interpretation, now standard, is inadequate in two ways. First, it is phenomenologically invalid insofar as it treats the contents of experience as stand-alone or discrete bits that can be isolated from their holistic subjective context. Second, and more important for the moment, the standard conception of the NCC obscures the actual nature of the empirical research by framing it in the metaphysically neutral language of "correlates" of conscious contents rather than as part of a larger causal/mechanical explanatory strategy. A notable claim on behalf of the correlate idea is that the neutral language frees us from philosophical disputes over the mind/body relation, allowing the science to move independently (Crick, 1996, p.486). Opponents of the standard NCC concept argue that the new neuroscience of consciousness needs philosophy now more than ever (Noe & Thompson, 2004). Certainly philosophy is still important in this context. But what is needed now is not so much more philosophy of mind but better philosophy of science.

The first issue facing the standard NCC concept concerns the nature of conscious content and the individuation of conscious states. This debate in philosophy of mind has been the primary focus of the NCC literature. The second issue has received less attention. It concerns the proper interpretation of the research itself, and thus it is an issue in philosophy of science. While it remains true that the neuroscience by itself does not adjudicate conceptual disputes regarding the nature of mental content, the experimental paradigms and explanatory canons of neuroscience are <u>not</u> neutral about the mechanical relation between consciousness and the brain. In what follows I consider both issues confronting the standard NCC: its assumptions about consciousness and its assumptions about neuroscience. I first introduce the idea of an NCC as presented by Chalmers (2000) and rehearse the phenomenological criticism of it, focusing mostly on the version formulated by Noe & Thompson (2004). I then argue that the neurobiological research in view is best characterized as an attempt to locate a causally relevant neural mechanism and not as an effort to identify a discrete neural representation, the content of which correlates with some actual experience. It might be said that the first C in "NCC" should stand for "causes" rather than "correlates." I conclude by attempting a revised definition of the NCC, aimed at connecting it with a mechanistic philosophy of neuroscience and avoiding the phenomenological problems associated with the standard version.

1 The standard approach to the NCC

David Chalmers (2000) describes a research program in neuroscience dedicated to the search for the *neural correlates of consciousness* or *NCC*. Chalmers defined an NCC as follows:

> An NCC is a minimal neural system N such that there is a mapping from states of N to states of consciousness, where a given state of N is sufficient, under conditions C, for the corresponding state of consciousness. (Chalmers, 2000, p.31)

Chalmers argues that this conception of the NCC is appropriate given what we know about consciousness and the brain, and that the goal of identifying NCCs can be attained in the foreseeable future (Chalmers, 2000, p.38).

Following Chalmers, I will focus on a particular kind of research aimed at a particular kind of consciousness, which he calls *content consciousness*. Content consciousness is primarily distinguished from *creature consciousness*, which is a property of a whole animal. Creature consciousness is the property of being conscious, as opposed to being unconscious as in a coma. In contrast, the search for the NCC for content consciousness involves finding brain differences that correlate with consciousness of particular intentional objects, such as visual stimuli. This research compares states of the brain in which the subject is unconscious of the stimulus with states of the brain in which the stimulus is consciously experienced. The paradigmatic example of research that aims at discovering the NCC for content consciousness is a set of experiments conducted by Leopold & Logothetis (1996) and Sheinberg & Logothetis (1997).[1] I pause to describe this research.

1.1 The research paradigm

Logothetis' primate laboratory was the site of a series of experiments that exploit the phenomenon of binocular rivalry. In binocular rivalry, competing stimuli are presented, one to each eye. The result in consciousness is an oscillation between perception of one stimulus and perception of the other. The subject does not perceive a hybrid composed of both stimuli, but instead experiences a flipping back and forth between the two. As the name "rivalry" suggests, there is a sort of competition between neural processes, in which the winner takes all. Thus, as one or the other of the visual stimuli dominates neural activity, we become conscious of it. With the use of electroencephalographic recordings, Sheinberg & Logothetis (1997) first identified neurons in the visual system of a rhesus macaque that respond to particular visual stimuli, such as a horizontal grate or a sunburst. They then trained the macaque subject to report its visual experience by pushing a button upon seeing the preferred stimulus pattern. The stimuli were then presented in a binocular rivalry situation, so that the monkey would respond when the preferred stimulus became dominant. Thus, the experimenters had a way to compare neural activation across very specific states of content consciousness – one in which the preferred stimulus is not conscious because suppressed by the rivalry, and one in which it is the "winner." The point to emphasize here is that binocular rivalry is form of intervention in the visual system, such that when the perception of the preferred stimulus is absent, this is explained by the mechanism of suppression. The experimental paradigm, then, does <u>not</u> merely display a "correlation" between two variables, one neural and the other conscious. It is employed within a specific and well-established causal/mechanical framework for neurobiological research. I return to this point in part 3 below.

The elegance of the paradigm lies in its ability to measure the difference in brain state that underlies changing consciousness of the stimulus, even while stimulus presentation is held constant. With this experimental design, these researchers have found a way to work directly in the field that Fechner called *inner psychophysics*. Inner psychophysics is the empirical study of the relation between the conscious mind and neurophysiology. Fechner distinguished inner psychophysics from *outer psychophysics*, which is the study of the relation between the mind and external stimuli. Historically, inner psychophysics had to be studied indirectly through outer psychophysics, because neuropsychologists could not precisely manipulate and measure neural activity independently of external stimulation (Hilgard, 1987, Boring, 1950). The binocular rivalry paradigm, together with improved microelectrical recording techniques, allows limited but quite direct experimental manipulation of the inner psychophysical relation. The results indicate that there is a significant relation between the winning visual perception and activation of particular neural populations in the inferior temporal cortex (IT) and the superior temporal sulcus (STS). Whenever the monkey reported seeing the preferred stimulus, 90% of the pre-identified neurons recorded in IT and STS were strongly active. However, this activity was almost extinguished when the monkey saw the ineffective stimulus (Sheinberg & Logothetis, 1997, p.3413). Thus, these seem to be the neurons that make the difference for consciousness of that visual content. Chalmers identified this work as being paradigmatic of the NCC research program, and attempted to derive his general definition of the neural correlate from it.

2 The neural correlate of what?

Does the content of subsystem N specify an actual conscious content? I now turn to a phenomenological critique of the standard NCC concept just introduced and illustrated. The basic theme is well known: meaningful experience at a given time is unified in a way that precludes the individuation of component "experiences," each of which exists on its own and carries an intelligible content. Recent book-length treatments of this theme include Michael Tye's *Consciousness and Persons* (2005) and Barry Dainton's *Stream of Consciousness* (2000). Returning to the passage from Tye quoted in Chapter 2:

> ...there really are no such entities as purely visual experiences or purely auditory experiences or purely olfactory experiences in normal, everyday consciousness. Where there is phenomenological unity across sense modalities, sense-specific experiences do not exist. They are figments of philosophers' and psychologists' imaginations. And there is no problem, thus, of unifying these experiences. There are no experiences to be unified. Likewise within each sense: There are not many simultaneous visual experiences, for example, combined together to form a complex visual experience. There is a

single mulitmodal experience, describable in more or less rich ways. (Tye, 2005, p.28)

Tye goes on to argue, among other things, that this phenomenological analysis is compatible with the empirical facts about the distributed and piecemeal nature of the underlying neural processing. This is exactly right, and I will return to this point at the end of the paper. However, this holism of experience is not compatible with the NCC for content consciousness as Chalmers defines it.

According to Chalmers' definition of the NCC, the neural subsystem N will be sufficient for the occurrence of the conscious state, under conditions C. The NCC will be activation in the smallest ("minimal") neural system such that, when you have the neural activity you have the content consciousness. Or again, the NCC will be the neural state that is minimally sufficient for the corresponding conscious content (Chalmers, 2000, pp.24–5). Call this aspect of the definition of the NCC the *sufficiency requirement*. Since there might be more than one neural subsystem that correlates with a given type of conscious content, and since these multiple neural states might not correlate with one another, Chalmers holds that the notion of a neural correlate should be understood to refer to a neural state that suffices for a given conscious state but is not necessary for it (Chalmers, 2000, p.24).[2] From a strictly logical standpoint this is entirely reasonable, though discussion in part 3 below will indicate that it is incomplete and misleading as a description of the experimental strategy. But before exploring that point further I first discuss the phenomenological validity of the standard NCC definition and the sufficiency requirement it stipulates.

The sufficiency requirement for neural correlates seems misplaced when considered from a phenomenological point of view. The ceteris paribus clause ("under conditions C") indicates that the minimal system is sufficient for the conscious content only on condition that everything else is in place *including the further neural conditions for consciousness itself* – creature consciousness, background consciousness, and other conscious contents to which it is related (Searle, 2004, Hohwy, 2009). Whenever these conditions obtain, N is held to be sufficient for "an experience," the content of which is specified by the activation profiles of N. But phenomenologically, this is an abstraction. "The" experience individuated in this way does not exist as such, but only as part of a global consciousness that clearly does *not* correlate with N. That is, there is no meaningful experience that stands alone and has exactly the content attributed to N. Clearly something is missing from the NCC if it doesn't actually correlate with any experience. Intuitively, one might construe the identified neural activity simply as a contributor to the genesis of a global conscious experience ultimately having as one component the particular visual feature to which these cells are attuned (say, horizontal orientation of the grating). But in that case these momentary activations in local neural populations in visual system, which the definition of the NCC is carefully constructed to pick out, are not sufficient for any actual experience had by any actual subject. What is missing? The beginning of an answer lies in the fact that consciousness of a particular visual content is always consciousness for some subject. The global features of this first-person point of view are part of the experience itself but not part of the NCC as defined by Chalmers. Noe & Thompson (2004) have considered this problem for the NCC concept in some detail. They argue that the contents of visual consciousness have certain global features that cannot be pinned to the firing rates of neurons in specific areas in the visual system. That is, they hold that the experience and the putative neural correlates do not have one and the same content (Noe & Thompson, 2004, p.13). Noe & Thompson list three features of experiential content not "matched" in neuronal receptive field representations. Visual perception is: *structurally coherent, intrinsically experiential and active* (Noe & Thompson, 2004, p.14). The authors conclude that the NCC program is a flawed approach to content consciousness.

2.1 Intrinsic experientiality and inner psychophysics

I focus on Noe & Thompson's second objection to the notion of an NCC, that consciousness is intrinsically experiential.[3] What Noe & Thompson have in mind here is roughly what various philosophers have called *subjectivity* or the *first-person point of view*. They write: "Perceptual content is intrinsically *experiential*, in the sense that the content of an experience is always the content as represented from a point of view" (Noe & Thompson, 2004, p.16). Similarly, the subjective point of view may be said to be an internal structure of consciousness, in the sense that what-it-is-like is always experienced "from the inside." This means that the first-person perspective is partly constitutive of content consciousness itself. One immediate consequence of this is that we cannot just stipulate, as Chalmers does, that we are studying content consciousness rather than creature consciousness since the latter is an internal structure of the former. Noe & Thompson clarify this point by explaining that their notion of experientiality refers to the way in which experience is *perspectivally self-conscious*. This means that "the perceiver and the world enter into the content of experience as background conditions"

(Noe & Thompson, 2004b, p.90, 91fn). So the content of experience is always perspectival, and part of the experiential content is *that* it is perspectival. This perspectival content concerns various possibilities for action from this vantage point, as well as expectations about changes in the point of view that will result. In this way, subjective, first-person experience is apprehended as for me, in a way such that "I" am specified by the structure of the experience itself (see Chapter 2). So, when Noe & Thompson criticize the NCC program for not taking into account the intrinsically experiential nature of conscious content, they are referring to the well-established phenomenological analysis of the subjectivity of first-person experience.

Note, however, that this phenomenological argument is not directed against neurobiological research on consciousness in general. It only concerns Chalmers' assumption that the holistically structured contents of consciousness will match or correlate directly with particular receptive field contents in isolated neural populations. Noe & Thompson call this faulty assumption the *matching content* assumption. The matching content assumption is expressed in the definition of the NCC as the requirement that the correlate be sufficient for the contents of visual experience.[4] This sufficiency requirement was discussed above.

It must be re-emphasized that the intrinsic experientiality of the content of consciousness does not just consist in creature consciousness. It is not just the fact that the subject is awake. Instead, the problem arises from the way the standard NCC conception attempts to individuate conscious contents according to momentary sensory states. There is no individual experience, the content of which is exhausted by the appearance of the preferred stimulus. Instead, the preferred stimulus is nested in an intelligible way within the ongoing global experience of a conscious subject. The result is that it remains indeterminate which conscious content is specified by the activation profile of N (and which is not specified). Partly on account of this, Noe & Thompson offer the notion of an *agreement* in content as opposed to a match (2007, p.11). Two representations can agree without having identical content. The statement "birds are in the sky" agrees with a photo of birds in the sky, but does not match it. The latter also represents a specific number of birds, their spatial relation to each other, their direction of flight, etc. So although the content of the minimal subsystem N lacks the subjective content of the conscious experience, Noe & Thompson concede that it does agree with the conscious content. I follow them here, but prefer to say simply that the momentary local neural activation underdetermines the content of consciousness. It is compatible with multiple conscious meanings, but does not specify any one of them. In the rivalry experiments, this can be illustrated by asking whether subsystem N specifies the conscious content "horizontal grating," "horizontal grating oddly oscillating," "horizontal grating like I have never seen a horizontal grating," "hallucination of a horizontal grating," etc.

2.2 Responses to the critique

At this point the proponent of the standard definition of the NCC invokes a distinction between the *total* NCC and the *core* NCC (Chalmers, conference correspondence). Chalmers draws this distinction in his paper, noting that it is adapted from Shoemaker's (1981) distinction between the total and core realization of a functional mental state. That distinction was not originally intended to apply to conscious content but rather to cognitive representational content. However, Chalmers directly transposes Shoemaker's idea of a core realizer of representational content to that of the core NCC, and writes without further ado:

> A total NCC builds in everything and thus automatically suffices for the corresponding conscious states. A core NCC, on the other hand, contains only the "core" processes that correlate with consciousness. The rest of the total NCC will be relegated to some sort of background conditions, required for the correct functioning of the core. (2000, p.26)

Here two ideas are used interchangeably, that of an NCC for content consciousness and that of a core NCC. The assumption is that perception of the preferred stimulus constitutes "the" core experience, and that there will be a discrete representational state of subsystem N that carries the content of this core experience. However, the phenomenological critique rehearsed above shows that the idea of a core experience is an abstraction. This means that the core NCC, which is now offered as the fruits of the binocular rivalry paradigm, does not correspond to any actual experience but only to a cognitive content which is said to be *at the core* of the holistic experience that in fact only corresponds to the total NCC. But again, the content individuated in this way is not the content as it appears to the subject in actual experience. Since a neural correlate ought to be the correlate of some experience, and there is no core experience for which the core NCC is minimally sufficient, it is useless to search for a core NCC in this way.[5]

Metzinger (2004) similarly responds to the phenomenological critique of the NCC for content consciousness by retreating to the total NCC (also see note 6). He concedes that the content of subsystem N will not be sufficient to specify the content of any actual experience, maintaining that consciousness will be correlated only with global states of the brain and nervous system. But notice that this move carries a high price. Salvaging the NCC definition in this way invalidates the binocular rivalry paradigm itself! By hanging on to the sufficiency requirement but conceding that only global brain states are sufficient, this response throws out the baby but keeps the bathwater. The kind of experiment conducted by Logothetis et al, which was originally supposed to be the paradigm for the search for the content NCC, now looks completely misguided – the very paradigm of how not to find the NCC – because it picks out local neural populations not global correlates. This result is backwards. The problem is not with the experiments. The problem is with our theoretical understanding of them.

As I pointed out above, the binocular rivalry design is essentially a classical experiment in psychophysics. But the sufficiency requirement (matching content doctrine), stipulated by Chalmers, is an unnecessary theoretical accompaniment. Where does this leave us? In the final section I argue that the search for the NCC can be reconceptualized as the search for the working parts of a neural mechanism. Discussion will show that this interpretation basically conserves Chalmers' analysis of the abstract logical form of the data, but strengthens it in a way avoids the phenomenological critique.

2.3 The enactivist interpretation reconsidered, too

But before moving on, I must clarify that, although the causal interpretation of the NCC advocated below is consistent with the holism of conscious content in a way that the correlate interpretation is not, it also differs from Thompson & company's enactive approach in several ways (Noe & Thompson, 2004, Thompson & Varela, 2001, Thompson, 2007). In general, enactivists share with Chalmers (and most philosophers) the assumption that binocular rivalry data are best understood as correlation data. While Noe & Thompson argue that there are no neural correlates of consciousness, they write as though neuroscientists must be as committed to the idea of neural correlates as their philosophical apologists: "…the NCC research programme is deeply wedded to a problematic and controversial internalist conception of the content of perceptual experience. The moral to be drawn is that neuroscience, far from having freed itself of philosophy, needs the help of philosophy now more than ever." (Noe & Thompson, 2004, p. 26). But it is precisely the standard *philosophical* interpretation of NCC research that is wedded to the internalism that Noe & Thompson find objectionable, not the binocular rivalry experiments themselves. In this respect at least, Crick's (1996) claim to the neutrality of the research still stands. The debate about the nature of conscious content is, strictly speaking, external to the experimental practice. For an adequate interpretation of the research itself, one must leave the domain of philosophy of mind and enter into philosophy of science. And when one does this, the mechanistic nature of the research becomes clear.

Thompson & Varela (2001, p. 425) do briefly discuss the idea that there is a causal relation between conscious experience and large-scale neural dynamics, and conclude with a call for *future* research that is not simply correlative, but causal. But the focus of their discussion is on emergent or downward causation. Their primary concern is to motivate the idea that conscious states (provisionally identified with large-scale neural synchrony) can make a difference to local neural firing rates by entraining the activity of smaller neural populations within the larger ongoing dynamics of the brain. And while Thompson & Varela do allow for the theoretical possibility of "upward causation," in which smaller-scale neural events can effect large-scale neural dynamics, they nonetheless maintain that "the available evidence so far...is only correlative, not causal; there is still no direct proof that changes in synchrony lead to changes in either behavior or consciousness" (Thompson & Varela, 2001, p. 419). While it may be true that there is no evidence that changes *in large-scale synchrony* (changes of the sort that Thompson and company are interested in) cause changes in consciousness, the primary burden of the next section is to clarify the evidential situation and set the record straight: the binocular rivalry paradigm does indeed provide empirical evidence that changes to local firing rates in IT and STS cause changes in visual consciousness. Finally, Thompson argues not only for a holism of conscious content but also for a fairly strong holism of neural function. He suggests that the embodied brain cannot be decomposed into component parts that, taken together, explain the behavior of the system as a whole (Thompson, 2007, p. 423). But discussion will show that the binocular rivalry data can be understood through a "component mechanism" view of neurobiological explanation, in which properties of the organized whole can be explained in terms of the action and dynamic interaction of the component parts.

3 The search for causes

Recall from section 1: Chalmers holds that the search is for neural correlates rather than causes, that the notion of a correlate should be understood as the logical relation of minimal sufficiency, and that the binocular rivalry design is a paradigm of this research, providing just this sort of data. But the research can and should be interpreted in a different (though formally similar) way, as part of the hunt for causes (the phrase is from Cartwright, 2007). Something very much like the relation of minimal sufficiency in Chalmers' definition of neural correlates can be found in J.L. Mackie's (1980) account of causality, which he analyzed in terms of *inus conditions*. The inus condition is the *i*nsufficient but *n*ecessary component of an *u*nnecessary but *s*ufficient set of conditions. Thus, a lit match can be an inus condition for a forest fire, despite the fact that it is neither necessary nor sufficient for the forest fire. In the explanation of the actual fire, the lit match plays a special role within the context of a set of conditions that are jointly sufficient for the fire. "What we call a cause typically is, and is recognized as being, only a partial cause; it is what *makes the difference* in relation to some background or causal field." (Mackie, 1980,p. xi emphasis added). Causal claims based on the inus analysis are to be tested by constructing experimental conditions that manipulate the putative inus condition in order to observe what difference this makes in the system as a whole. And this is a very precise description of the experiments here in view. As discussed above, Logothetis et al identified the neural activations that *make the difference to*, and hence are partial causes of, the experience of the preferred stimulus.

It is important to realize that Mackie offered the inus condition only as the formal aspect of a larger account of causation, which he took to be an ontically fundamental relation. Mackie argued that the inus condition picks out a real cause (and not a mere logical correlate) only insofar as it identifies an *effective* relation (or one in which A "fixes" B, A "produces" B, etc.). Roughly, A effects B when <u>preventing</u> A prevents B. The binocular rivalry paradigm offers a way of testing precisely this: the rivaling stimulus intermittently *prevents* the neural activation (A), and in turn this is shown to *prevent* the behavioral report (B).[6] With this in mind, the research clearly suggests that activation in the neural subsystem N is *causally relevant* to the consciousness of the preferred stimulus. The experimenters observed what happens both when N is active and when it is *in*active. This suggests that they are treating subsystem N as an underlying mechanism rather than as a logically sufficient condition.[7]

Carl Craver (2007) offers a detailed account of the explanatory strategy in neuroscience that is very much in tune with the idea that the inus condition identifies causes in terms of their effectiveness.[8] Craver argues that explanation in neuroscience takes the form of a search for mechanisms that underlie cognitive and behavioral phenomena, and that these mechanical explanations are offered on the basis of evidence about *manipulability*. The basic idea behind Craver's "manipulationist" view is that X is *causally relevant* to Y if and only if there is an "ideal intervention on X that changes the value of Y, or the probability distribution over values of Y" (Craver, 2007, p. 198). He writes:

> An ideal intervention (I) on X with respect to Y is an intervention on X that changes the value of Y, if at all, only via a change in X. Controlled experiments are designed to rule out the possibility that the change in Y is due to the effect of variables other than I.... (Craver, 2007, p. 199)

Although the intervention achieved through binocular rivalry is far from ideal, it clearly displays this manipulationist strategy. The experiment manipulates the activity of specific neural populations deep within the visual system by setting up competing activity that suppresses those particular neurons.

If one puts aside puts aside the phenomenological issue discussed in part 2 above (whether the content of subsystem N specifies an actual conscious content) then Chalmers' vocabulary of minimal sufficiency, ceteris paribus conditions, and core/total NCCs can be made to pick out the same bare logical relation picked out by the inus condition. The total NCC will be unnecessary but sufficient for the state of consciousness, while the core NCC (i.e., subsystem N) will be insufficient but plausibly necessary for it (Chalmers, conference communication). Even so, the two are not quite equivalent. The language of minimal sufficiency implies that neurobiological research aims to turn up covering laws of the sort familiar from the tradition of logical empiricism. Thomas Metzinger, in the volume *Neural Correlates of Consciousness* which he edited and in which Chalmers' paper appeared, straightforwardly assumes that empirical claims about NCCs will be formulated as laws stating which conscious contents will follow "with nomological necessity" from the activation in subsystem N (Metzinger, 2000, p. 285). As discussed above, since the activation in N underdetermines the eventual conscious content, this claim cannot be sustained. But the present point is that wielding the sufficiency requirement in this way depicts scientific inquiry as the construction of deductive arguments that derive modal relations between observation statements. It is widely held in contemporary philosophy of science that biology and psychology don't work that way.[9] Craver, for his part, argues that "derivability" should be jettisoned as the goal of scientific explanation (2007, p. 160). Instead, both Craver's manipulationist approach and Mackie's inus analysis take it that the aim of most experimental science is to discover the causal structure of the world. Knowing the causal structure of a thing involves knowing how it works; i.e., being able to intervene in and manipulate the system. This does not require laws with nomological necessity.

Obviously, performing the sorts of neural manipulations required for the science of inner psychophysics is no easy task, which is why the rivalry design makes important progress. Imaging techniques can provide heuristic guidance for generating causal hypotheses, but in the absence of precise neural manipulations their explanatory value is quite limited. Three broad types of neural manipulation are used in animal studies: psychological, chemical, and electrical (Panksepp, 1998). The binocular rivalry paradigm is an example of the first variety, and it provides an unusually specific and local manipulation. But future research in the neural causes of consciousness may more often exploit the third kind of intervention. Transcranial magnetic stimulation (TMS) offers the prospect of direct and precise neural interventions that are minimally invasive. In TMS, a tiny magnetic field is projected onto the surface of the brain, depolarizing cortical neurons and modulating their activity (Bohning, 2000). Shutter, van Honk & Panksepp (2004) argue that TMS can be an important tool for researching the fine-grained causal structure of the neural networks in the cortex.[10] Thus, binocular rivalry studies are just one tool available for research on the neural *causes* of consciousness.

3.1 Two objections

At this point the causal/mechanical interpretation of NCC research will face objections from two quarters, supervenience theorists and enactivists. To address the first, a brief detour back into the metaphysics of mind is unavoidable.[11] It may be argued that brain states are the realizers of conscious states, and hence cannot be the causes of them. The basic response to this objection is that there is no contradiction in holding that causal mechanisms can be constitutive too, as component parts of the larger organized system that displays the explanandum. First, note that the gestalt structure of conscious content (discussed in part 2) means that subsystem N does not constitute a complete supervenience base for any conscious state. N is better thought of as a component in a system that realizes conscious experience. In the case of component mechanisms (as opposed to causal mechanisms that are not constituents of the system displaying the explanandum), there is an added requirement of *mutual manipulability*. Here, the activity of the <u>component</u> is also shown to be manipulable through an intervention on the larger systemic property it underlies (Craver, 2007). That is, if the conscious experience changes in the relevant way, then so will activation in N. As noted above, this is just what Logothetis et al demonstrated. But the component-to-system relation in view here is slightly different from the realization relation. In the latter relation, there will be no changes to the supervening phenomenon without some change in the realizing base. Supervenience, then, holds between ongoing, holistically individuated conscious experience and the <u>entire</u> organized system that realizes it. But the relation of mutual manipulability holds between a specified system level change and one of the components of the system that underlies it (Craver, 2007, p. 153 fn33). And again, this is different from a minimally sufficient correlate. On minimal sufficiency, a change in conscious state will <u>not</u> predict a change in N. Thus, a local mechanism such as subsystem N can be causally relevant when it is a mutually manipulable inus condition in the larger system that exhibits the mental property. In this way, the search for the NCC may be understood as a search for the neural components of a complex mechanism (an organism) which is conscious.[12]

Enactivists will object on somewhat different grounds. Following Maturana & Varela (1980), Thompson (2007) rejects the idea of a component neural mechanism advocated here. Paralleling his view of the holism of conscious content (which I accept, see part 2 above), Thompson suggests that the brain, too, must also be understood holistically as a non-decomposable system (Thompson, 2007, pp. 422–3).[13] Since the nervous system constitutes a self-organizing dynamic system that maintains own structure and its own boundaries, Thompson holds that there is no value in talk of "pre-existing" components discoverable through experimental manipulation or reverse engineering. Hence, he opposes localization studies in neuroscience because they do violence to the prior holism of the embodied creature. Any alteration or manipulation of the system creates a new unified whole, from which inferences about the original system cannot be made. Thompson quotes

Maturana & Varela here: "There is intrinsically no possibility of operational localization in the nervous system in the sense that no part of it can be deemed responsible for its operation as a closed network, or for the properties which an observer can detect in its operation as a unity" (Maturana & Varela, 1980, p. 129). This statement runs counter to the vast weight of practice in neuroscience, and to the "systems" tradition in philosophy of explanation, of which Craver's manipulationist view is a current example (see also Bechtel, 2008).

As discussed in Chapter 7, the basic characterization of the nervous system (and the organism as a whole) as an actively self-organizing system is both true and interesting, and it is certainly important to acknowledge the difficulties of inferring local function in any system with the degree of connectivity of the brain (a point also made by Chalmers, see note 2). But this does not entail that the nervous system constitutes a kind of monad without component parts. To investigate causally relevant components of a system is not necessarily to say that the components "pre-exist" the system as a whole, or that the behavior of the system is a straight linear sum of the action of its components. Even in an actively self-organizing system, the possibility remains for tracing system-level features back to the activity and interaction of the networked structures and processes that sustain them. A biological cell is a prototypical example of such a system in which various components can be said to be the mechanisms underlying activities of the cell as a whole. Although it is true that temporal dynamics have often been neglected in mechanistic philosophy of biology, there is no need to choose between mechanistic (bottom up) and complexity theoretic (top down) approaches (Bechtel & Abrahamsen, 2011).

Further, the causal/mechanical interpretation of NCC research advocated here is compatible with the holistic individuation of conscious states. The question of the *phenomenal unity* of conscious content is distinct from the question of the *neurophysiological unity* of the underlying mechanisms (Tye, 2005). The former does not require the latter. The fact that the different sense modalities have distinct neural processing streams, for example, does not entail that each processing stream constitutes its own "experience." Another way of putting this point would be to say that the *binding problem* is not the problem of how mini-experiences are bound together (in a Humean bundle) to create the illusion of unity. Rather, it is the problem of showing how various disunified neural processes give rise to one single experience at a given time. Thus, phenomenal unity is a matter of parallel processed sensory information entering into one and the same phenomenal content. Consider the example of an experience of a red square next to a green square. Surely information processing about the two squares is not neurophysiologically unified. Each square is "represented" on its own. But this does not show that there is more than one experience, each existing on its own. Even though representations are proper parts of an experience, this does not mean that they are themselves experiences (Tye, 2005, p. 40). If neural activity is distributed, parallel, and dynamic, then physiologically disparate processes can be part of a <u>single</u> mental process. There can be *mechanisms* without *modules* (Spivey, 2007, p. 28).

3.2 Neural mechanisms and the explanatory gap

One more important point must be clarified. I do not suggest that this research has the resources to bridge the so-called explanatory gap and solve the problem about qualia. I follow Levine (2001) in holding that we will not have a complete explanation of consciousness until we have a functional analysis of qualia per se. But even Levine recognizes that we *can* in fact locate the causes of consciousness, despite the absence of a functional-role analysis of qualia. We can have good empirical reason to believe that particular neural activations are what bring about particular features of experience. There is no conceptual problem in conducting this sort of investigation. What we are at a loss to do is to explain why these neural mechanisms (or any other) should give rise to the particular qualia they do (rather than some other other). Levine even suggests that it is precisely the fact that we can localize the mechanisms of consciousness that motivates the absent qualia and conceivability arguments in the first place! That is, the moment we identify c-fibers (say) as the mechanism of pain, we immediately find that we can imagine that particular mechanism working in the absence of painful qualia. So the explanatory gap argument is not a denial of the claim that c-fibers (say) are the causal mechanism for pain. Rather it is a denial of the claim that we understand why this should be the case, since we have no analysis that connects what c-fibers do with what pain feels like.

At hand is a very specific actual case that provides concrete illustration of the philosophical point about the explanatory gap. According to this data, N is the mechanism causing the change in the experience, but we certainly have no idea why this should be the case. A possible source of confusion here is that the concept of explanation that is in use in the explanatory gap arguments in philosophy of mind is very different from the concept of explanation in use in mechanistic neuroscience. On the former notion, to explain a thing is to make it intelligible; i.e., to show why it should be the case (Jackson, 1994. See Chapter 7). Doing this requires a logical derivation of the explanandum from the observation plus physical laws. But as discussion has already shown, the operative notion of explanation in neuroscience is weaker: to explain a thing is to know how it fits in the causal structure of the world. Both standards of explanation are perfectly legitimate, and it remains true that by the former standard there is an explanatory gap. But, by any standard, causal mechanisms are being investigated, not logical correlates.

4 Conclusion: the NCC Reconsidered

Phenomenological analysis shows that we cannot directly map abstracted bits of experienced content to momentary activation in isolated neural populations. Again, this is because actual experience is a holistic gestalt, not a patchwork of independent bits. There is no non-subjective, non-global bit of meaningful visual experience to which the activity in IT and STS could conceivably correspond. But there is a precisely measurable difference between the neural activity that underlies two distinct, globally subjective moments of conscious experience. Binocular rivalry research attempts to identify the neural activity that makes the difference between these moments. "Making the difference" is better analyzed in terms of causal relevance than as the specification of a conscious content. In turn, causal relevance is investigated by neuroscientists through the manipulation of a variable that can be formally described as an inus condition. Thus, the research aims at localizing the working parts of an underlying mechanism, and in this respect it is continuous with much of the rest of neuroscience as well as the long tradition of psychophysical research. Finally, a causal interpretation of the NCC avoids the phenomenological problems associated with the idea of a contentful correlate for a core experience.

In light of the above, I attempt the following of the definition of a Neural Cause of Consciousness, or NCC:

> An NCC is a minimal neural system N such that states of N are underlying causes of a measurable change in consciousness, where a given state of N, as the causally relevant component of an embodied mechanism, is a mutually manipulable inus condition for the specified aspect of the conscious state.

Postscript: Neurophilosophy, Darwinian Naturalism, and Subjectivity

A neurophilosophy of subjectivity is taking shape. "Neurophilosophy" as a trademarked proper name refers to the work of Patricia & Paul Churchland, but there is a whole array of contemporary research in social science and the humanities that may be designated "neurophilosophy" in a more generic sense. Bickle & Mandik (2010) distinguish between neurophilosophy and philosophy of neuroscience. The former is the application of neuroscience to philosophy while the latter concerns foundational issues within the neurosciences, i.e., concepts and methods of neuroscience. This distinction, while perfectly serviceable, is somewhat misleading. A central element of neurophilosophy is the "coevolutionary" approach, which seeks reflective equilibrium between neuroscience and philosophy. This method certainly entails doing some of what Bickle & Mandik label "philosophy of neuroscience" rather than neurophilosophy. I take the term to refer more generally to any work that is concerned with the wider conceptual significance (or lack thereof) of neurobiology for topics related to mind, person and society. Within this wider literature there is great inconsistency, disagreement, and divergence, and so neurophilosophy thus construed does not consist in any unified ontology or methodology. Large-scale currents can be discerned, however, including *neurophilosophy* proper (philosophy of mind, philosophy of science), *neuroethics* (moral psychology, meta-ethics, applied ethics), and *neurophenomenology* (consciousness, subjectivity, and embodiment).[1] While almost all this literature at least presents itself as naturalistic, there are many varying conceptions of what this means and differing attitudes toward science and "scientism." Here in the postscript a brief comment is in order regarding the place of this book within the wider neuro literature.

The neurophilosophy of subjectivity pursued here is a piece of Darwinian naturalism. According to the usual gloss on "naturalism," it is the idea that philosophy should be methodologically continuous with natural science. On this Quinean formula, the general goal for philosophy is to be informed by science whenever relevant, which is far more often than philosophers have traditionally allowed. The specifically *Darwinian* naturalist thinks that biology is an especially robust resource for extending the reach and relevance of natural science beyond its standard preserve (cf. Lewens, 2005). For the science of subjectivity, the historical character of Darwinism is the element otherwise missing from philosophy of mind and cognitive science. The point isn't merely that the causal explanation of subjectivity will ultimately advert to evolutionary theory. The deeper implication is for the concept of subjectivity itself, the explanandum not just the explanation. Darwinism brings a genealogical perspective, which can provide critical purchase on traditional humanist ideas about the nature of experience.[2] Subjectivity is a legacy of the profoundly theological worldview of the Enlightenment and Romantic periods (Davies, 2009). As I suggested in the introduction to Part Two (Chapter 5), subjectivity belongs to what Sellars (1963) dubbed the *manifest image* of mind and person, and Darwinian naturalism suggests that this manifest image is in serious trouble (Flanagan, 2003).

Darwinism extends the reach of natural science into the manifest image generally, and to individual subjectivity in particular, by bringing genealogical tools to bear on the categories and intuitions that guide the way we conceive of ourselves. The putative explanatory limits of the scientific image were fixed by these conceptions about our nature and about what needs to be explained (i.e., they have been fixed in relation to the manifest image). Genealogy shows how ideas such as subjectivity are historically conditioned, and thus how they are open to revision. But this result must be handled with care. It is not a simple matter of announcing that the first-person perspective was selected for – a proclamation which could then be used to dismiss other cultural, linguistic, political, phenomenological or conceptual analyses. Darwinian genealogy is not a trump card. It is an additional resource for critique, one that can lead astray as easily as any other.

1 Constraints on neurophilosophy

Any naturalistic philosophy must somehow navigate between the opposing threats posed by the *naturalistic fallacy* and the *genetic fallacy*. The naturalistic fallacy – that of inferring what should be from what is – is the better known of the two mistakes. At a basic level, this mistake is an expression of a general human tendency to forget critique. In the context of neurophilosophy, the danger of the naturalistic fallacy is to suppose that facts about the brain justify some practical or cognitive norm. For example, the fact that human decision-making is often governed by quick and dirty emotional responses might be mistaken for evidence against the quaint humanistic idea that decisions *ought* to be taken according to one or another abstract principle. But, as Moore put it in the Open Question Argument, given any description of the facts, it always remains pertinent to ask, "Is it good?" Thus, it remains an open question whether quick and dirty emotional responses are really the best basis for decision-making, and whether this human foible might be modulated. Scholars in the humanities are permanently suspicious that every natural scientist is forever committing this fallacy. And indeed people often fail to be vigilant, forgetting the critical component of any investigation into normativity or value. Nevertheless, vigilance really *does* work; it is possible to be a naturalist without committing the naturalistic fallacy. With care, the mistake can be avoided by remembering that natural science by itself doesn't carry normative implication.[3] The hope that science might serve as adequate justification for normative judgments – or the pretense that there are no critical problems about norms – marks any writer as "naturalistic" in the fallacious sense.[4]

But for the Darwinian naturalist, the genetic fallacy is more interesting than the naturalistic fallacy. Roughly, the genetic fallacy is the mistake of concluding that a proposition must be false just from an explanation of how someone came to hold it (Sober, 1998). Now, naturalistic philosophers are often concerned to *debunk*. Their goal is to show that some received idea, entrenched norm, or dominant ideology is false or misleading. Genealogical arguments are then brought to bear, showing that the target is held in place by causes that are not rationally or epistemically related to the belief or ideology in question. Thus, Nietzsche argued that Christian morality descends from certain historical circumstances, and is nourished by particular psychological tendencies which have an independent causal/historical explanation. But though this genealogy be correct, it does not *by itself* show that Christian morality ought to be discarded. To assume that it does show this is to commit the genetic fallacy. A neuroethical example: if it is true that moral deliberation is partial to family and friends *only* because of affective association mediated by the amygdala, this does not immediately show it wrong to be partial to family and friends in this way (cf. Greene, 2003). The genetic fallacy, it may be seen, is a form of ad hominem. It attacks the holder of the judgment, not the judgment itself. This sort of attack can be simply irrelevant. Any genealogical argument is constrained to respect this limitation on what can be concluded about the content of belief just from the causal/historical explanation of belief formation and fixation. But though it is true that such arguments cannot *directly* establish the falsity of the target ideas, they can provide a great deal of indirect evidence about them. In particular, genealogy can show that the traditional and intuitive arguments in favor of received notions are not as robust as they may appear. Knowing the facts about belief formation and fixation can "prompt" us to reevaluate our intuitions and to reform our understanding (Greene, 2003). Wielded correctly, genealogy can be a powerful tool for critique. As will be seen in a moment, contemporary naturalists like Davies (2009) use genealogy in ways that closely parallel the Nietzschean exemplar.

2 Subjectivity: a dubious concept

Philosophers of a certain "tough-minded" temperament (as identified by William James (1995 [1907])) will wonder whether subjectivity might simply be a dead idea. Detecting a distinct whiff of sentimentality and old fish, they will suspect that subjectivity may be a relic best left to the history of ideas, like Platonic Forms or the Leibnizian Monad. Paul Davies is one such tough-minded Darwinian who warns that "naturalization" is not always as progressive as it appears. When a concept clearly comes down to us from sources located within an overarching theological worldview, he argues, we should regard that idea with suspicion (2009, p. 25). He calls such ideas *dubious by descent*. Unquestionably, subjectivity is dubious by descent. Alongside ideas about teleology, free will, and the soul, subjectivity was close to the heart of Enlightenment and Romantic metaphysics of the 18th and 19th centuries (Reed, 1997; Hacking, 1998). We know that these concepts stem from a profoundly non-naturalistic worldview, and this warrants pessimism about whether they can be salvaged today.[5]

Davies characterizes the general project of naturalization, found in much contemporary philosophy, as one of *concept location*. The concept location project is the attempt to find a "place" for received philosophical ideas within the worldview of contemporary science. The aim is "to integrate apparently important concepts from our humanistic tradition with the concepts and claims of our best sciences" (Davies 2014, pp. 76–77). To locate humanistic concept x is to tie x to some well-understood explanatory framework, such that the relation between x and the natural sciences becomes intelligible.[6] The concept location project, then, is the attempt to reconcile the philosophy of human nature with the sciences, and Davies sees this as a misguided effort to *save* the traditional worldview from the shock of the new. Cherished ideas return over and over, Davies argues, "found" in today's best empirical theories only because they are put there again by unreconstructed philosophers. Davies identifies two general domains in which the concept location project is flourishing: teleology and free will/responsibility.

Davies argues that the concept location project should be abandoned, and advances the following methodological precept for true Darwinian naturalists:

> Concept Location Project (CL): For any concept dubious by descent, expect that the concept location project will fail; expect, that is, that the dubious elements of the traditional concept will face revision or elimination as we analyze inward and synthesize across the concepts and claims of all the relevant contemporary sciences. (2009, p. 228)

Everything we have learned about the natural world and our place in it should lead naturalists to expect that cherished traditional concepts will *not* be rehabilitated. Our best Darwinian vision of ourselves is one in which we humans are psychologically biased and culturally conditioned to think in certain misleading ways, and we must constantly struggle to overcome these tendencies. Thus we should expect that traditional notions like self, purpose, and freedom will turn out to be radically mistaken. There will be no place – no "location" – for these ideas in a properly naturalistic worldview. In short, humanism will not be redeemed. Instead, these cherished ideas should be explained away as illusions generated by our deeply animal nature.

So, Davies' Directive (CL) indicates that concept localization is not the way to naturalism. The attempt to save these dubious ideas will at best produce a garbled and incoherent philosophy. Worse, the project of concept location can be positively damaging when it serves as cover for deeply reactionary doctrines by giving an air of progressive respectability to otherwise discredited and authoritarian crypto-theology (2009, p.26 and elsewhere). While some of Davies' more ominous political overtones might be debated, the general diagnosis rings true for much recent literature on subjectivity. For example, some research in "naturalized phenomenology" is ripe for this sort of criticism (e.g., Petitot, Varela, et al, 1990). "Enactivism," too, often deserves the pseudo-naturalistic label (e.g., Hutto & Myin, 2012. See also Thompson, 2007).[7]

Broadly speaking, all this literature is part of the philosophical tradition of "embodiment." At least one major strand of that tradition flows from Friedrich Schleiermacher, the father of modern liberal theology (*The Christian Faith* 1999, [1830]). More generally, the Scholastic tradition of Personalism exerts considerable undertow in philosophy of mind, and phenomenology.

Since subjectivity is a dubious concept, shouldn't we expect it to be eliminated? No. But we should expect it to be revised. CL predicts that "dubious elements of the traditional concept will face revision or elimination" (p. 228). Re-reading Davies, one gets the feeling that the possibility of revision (rather than elimination) is only grudgingly conceded. Including it renders CL significantly weaker, for now the directive only applies only to traditional concepts *in their received, Romantic or Enlightenment form*. Revised versions of these concepts, purged of their dogmatically anti-naturalistic elements, can be legitimately located within the Darwinian worldview. In fact, Davies argues that this is precisely what Darwin himself was up to! Darwin revised the concept "selection," for example, to purge it of teleological baggage (Davies, 2014, pp. 7–13). Natural selection is thoroughly non-teleological but this does not prevent it from being a naturalized version of the concept of "selection." Similarly, the "subjectivity" investigated in this book not the transcendental subjectivity of German Idealism and transcendental phenomenology, it is not the emergent quasi-vitalist autopoieisis found in enactivism, and it is not the direct inner awareness advocated by qualophiles. It is not even existential pre-reflective self-consciousness, though this is much closer. It is a newly revised notion, motivated by the life sciences, investigated by the cognitive sciences, deconstructed by the historical social sciences, and analyzed through the logic the first-person pronoun.

So, far from running afoul of Directive CL by pursuing a bankrupt project of concept location, the project of this book conforms to CL very closely. Davies' useful directive indicates that naturalistic revision of dubious concepts will take place as we "analyze inward" and "synthesize across" relevant sciences. Both of these things have been attempted here. First, to analyze inward is to find underlying mechanisms and functional decompositions. I argued in Part Two that the underlying mechanisms of the first-person perspective are affective-cognitive maps in the mesolimbic system. Second, to synthesize across the concepts and claims of relevant contemporary sciences is to render the concept continuous with those of the neighboring sciences. In classic naturalist form, this means something less than providing a logical derivation but more than just showing logical consistency. Basically, the revised concept should *mesh* with the relevant science in a way that produces the best big picture currently available (see Chapter 3 for more on "meshing"). I argued throughout the book for a notion of subjectivity synthesized across relevant concepts from philosophy, cognitive science, affective neuroscience, evolutionary developmental biology, Developmental System Theory, and others. Thus, this book offers a significantly revised and newly "relocated" version of the traditional philosophical concept of subjectivity, and I take it that this is what neurophilosophy should be all about.

But this kind of neurophilosophy differs subtly from vintage eliminativism. Consider that the old eliminativist rallying cry – *reduce or eliminate!* – lacks the genealogical bent I have emphasized. By identifying specific ideas that are dubious by descent, neurophilosophers can extend the reach of neurobiology into the humanities in a nuanced way. Adding this historical dimension can improve neurophilosophy even where significant progress has been made under the flag of eliminativism. For example, Mandik (2009) is an eliminativist about subjectivity, which he construes as *one-way knowability*.[8] Mandik argues for the elimination of subjectivity because there is really no good reason to think that life-lessons can *only* be learned first-hand. On the contrary, our best current theory suggests that neural networks could conceivably be configured by means other than experiential learning. Mandik's eliminativism is productive and interesting because he questions the received doctrine that subjectivity brings privileged knowledge via immediate inner awareness. But one-way knowablility is an element of the received notion of subjectivity that is dubious by descent. I argued in Chapter 4 that it descends from the Berkeleyan tradition of empiricism. By instead treating subjectivity as the identification-free structure of the first-person perspective, the idea of a subjective mode of presentation can be revised rather than eliminated. Further, the intuition about one-way knowability that sustains the Knowledge Argument (i.e., the intuition about Mary) can be reinterpreted in light of the revised notion. What experience teaches always comes in a specific form and always bears the tell-tale signs of a specific mode of acquisition, a particular history (cf. Lewis, 1990). To know a thing first-hand is to gain information about it by a distinctive route. So although Mandik is probably right that, in principle, information with the distinctive for-me structure could be directly hand-coded into the network, this is no grounds for eliminativism about the subjective mode per se. For, once the hand-coding is done, a first-person perspective has been engineered. Rather than eliminating subjectivity, then, we can demystify it.

So genealogy helps Darwinian neurophilosophers *make contact* with the manifest image. Without it, thinkers in both images will simply persist in the belief that their concepts exclude the possibility of consilience, and eliminativism will *always* seem the best option to a neurophilosopher. Genealogy is a critical tool that can be shared across the images to improve the frameworks of both. Genealogy was originally wielded as a form of internal critique within the manifest image. It was a way for those tough-minded historians, social theorists, and philosophers to gain critical purchase on the dominant ideas of the humanist tradition. But here is a tool which even tough-minded naturalists within the scientific image can draw on to improve their own practice. Neurophilosophers should employ it to help mediate between the two frameworks, to generate a wider view of the total terrain. This wider view, this synoptic and stereoscopic vision encompassing both the scientific and the manifest, was what Sellars proposed as *the* task of philosophy.

3 A sense of where you are

The Sellarsian conception of the aim of philosophy, as I noted in the brief introduction to Part Two, is to understand how things in the broadest sense hang together in the broadest sense. In this postscript I have been looking at the project of this book in naturalistic perspective, arguing that Darwinian genealogy is a key tool for the enterprise of seeing how it all hangs together. But the idea of synoptic interpretive activity also has resonance for the *topic* of the book, for the account of subjectivity per se. In the neurobiological image, subjectivity just *is* the way things hang together, for us, generating a subject position. Subjectivity is your sense of where you are, which is a sort of synoptic vision. The fundamental element of subjective mental life is not speaking or writing but *reckoning*. In turn, reckoning is a matter of seeing how things hang together. Operating at several levels of remove, philosophy is simply a reflective form of subjectivity.

But, unlike philosophy or other reflective kinds of speaking and writing, basic subjectivity does not require a "you" in any rich mental or psychological sense, whatsoever. This seems counterintuitive; how can I cope without conceptualizing myself? But counterintuitive as it may be, this points the way out of the maze of self-consciousness. Subjectivity is fundamentally relational; it is a matter of *difference* not identity. The first-person is composed of the differential relations among a set of interests and affective trajectories. The key realization is that the subject position is just that – a position. This *positionality*, well known to continental philosophers (Alcoff, 1988; Deleuze, 1994), is defined by its relation to a salience landscape. It is the perspective from which the landscape has precisely *this* salience. Thus concepts such as "I" and "me" are not constitutive of first-person subjectivity. All that is required is that things hang together, loosely indexed to "here" – a set of parametric values that form a trajectory in and around the environment. Embodied subjectivity is an affectively charged "place" which partly constitutes conscious experience. In the first instance then, subjectivity is not self-conscious.

The hanging-together idea is pretty obviously informal. The synoptic goal of the philosopher is not to unify the two worldviews, or to derive one from the other, but to *juggle* them, to coordinate them in a wider cognitive activity. According to Sellars, the manifest and scientific images will hang together "stereoscopically," not modally. Likewise, the unity of the first-person perspective is only loose and approximate, a practical gestalt, not a logical form. We constantly juggle various forms of salience, such that a stereoscopic first-person perspective is dynamically maintained. The various kinds of salience for-me need not be governed by any unitary objective norm, and they need not define any internally coherent "me." They need only hang together as I learn my way around. So, this is not another book about the self. The self is another sister in the family of manifest concepts that includes "person" and "soul." But the nature and existence of the self is not at issue in this book, and I remain officially agnostic about it. In its simplest form, subjectivity can happen in the absence of anything like a self, self-concept, or persisting person. It can even take place without much of a *subject* to speak of. All that is really required is a certain kind of animal – an animal like us – exploring its world. In the words of Elizabeth Anscombe, "These conceptions are subjectless. That is, they do not involve the connection of what is understood by a predicate with a distinctly conceived subject. The (deeply rooted) grammatical illusion of a subject is what generates all the errors which we have been considering" (1975, p. 65).

Notes

2 Subjectivity and Reference

1. Of course, this is a very incomplete list, naming only a few of the recent contributions salient to the present formula. Of the last century and more, the structure of the first-person has also been given extensive analysis by philosophers on both sides of the Atlantic including Husserl, Wittgenstein, Anscombe, Heidegger, Sartre, Merleau-Ponty, Shoemaker, Evans, and Santayana.
2. Bermudez (1998) and Evans (1982).
3. That is, no claim is made about which person is the subject (Pryor, 1999). See below pp#.
4. Campbell (1999), p. 90. The example that follows is also adapted from Campbell.
5. Pryor (1999).
6. It is also said that Berra once quipped, on receipt of a check in payment for appearing on Jack Buck's radio show, "How long have you known me and you still can't spell my name?" The check was made out to "Bearer." this nicely points to the two varieties of singular reference under discussion, by name and by index.
7. Nor can this be finessed by appeal to a reflexive name of the "thinker of this thought" variety. See the discussion of Rovane's Alternative, below.
8. Note that I extract the notion of identification-free self-reference from the rest of Evans' account, which employs a more properly semantic approach to I-thoughts. In particular, I do not take on board Evans' commitment to the so-called *Generality Constraint*. See discussion below pp. 17–20.
9. This analysis is quite different from the *inner awareness* approach to subjectivity championed by, e.g., Kriegel, 2009; Levine, 2001. I take up their views in Chapters 3 and 4, respectively.
10. Cf. Ismael (2007).
11. Evans, (1982), p. 170. Evans sometimes calls this "egocentric" thinking. He recognizes that Husserl and other phenomenological philosophers have given similar analyses of the structure of experience in terms of a "zero point" of perception. For my purposes, it is important to note that the characterization of subjectivity as "here" is a metonym for the first-person perspective. Baseline subjectivity is not only structured by a spatial "here" and a temporal "now" but also by situated affective valences, engagements, and motivations. More below.
12. Note that Campbell does not share this view. Instead he follows Evans more faithfully on this point, holding that "what makes an experience and experience of x's is the possibility of self-ascription of it by x" (p. 97). I avoid this formula because it may imply that animals without a self-concept (and hence without the power of self-ascription) cannot have experiences. Whether the Evans-Campbell formula carries this implication depends on what the "possibility of self-ascription" consists in. See below.

13. This only works for simple perceptual demonstration in which reference to the object is fixed by its location relative to the perceiver (construed suitably broadly to include affective "location" in the environment of concern). But perceptual judgments that do identify by description are also common, not displaying immunity. Campbell (1999)
14. See Tomasello (2014), etc.
15. For the record, I would opt for something like Campbell's "simple rule" account of the first-person pronoun.
16. O'Brien characterizes Evans' view as *neo-Fregean*. Ref.
17. Cf. Ismael, 2007, p. 73.
18. Following Evans, Gillett (1987) answers yes. Dreyfus (repeatedly) answers no. For a recent volume dedicated to this debate, see Schear (2013), *Mind, Reason, and Being-in-the-World: The McDowell-Dreyfus Debate*.
19. The Soderbergh film *The Girlfriend Experience* (2009) attempts to dramatize the holistic subtleties of subjectivity, intersubjectivity, and alienation when they become "lifestyle" commodities.
20. Further diverse examples, see Tononi (2012), Bayne & Chalmers (2003), and Dainton (2000). In chapter 8 it is further emphasized that subjectivity does not just consist in "creature consciousness" (Chalmers, 2000). That is, it does not just consist in the fact that the subject is awake. Instead, the content of experience is always perspectival, and part of the experiential content is *that* it is perspectival. This perspectival content concerns various possibilities for action from this vantage point, expectations about changes in the point of view that will result, along with affective valences and salience markers. In short, "...the perceiver and the world enter into the content of experience as background conditions" (Noe & Thompson, 2004b, pp. 90, 91 fn). And again, remember that the "perspective" here is not just spatiotemporal but also affectively engaged.
21. Of course, there are experiences of detachment and externality, feelings of alienation from one's own world or consciousness. But these take place *within* experience.
22. Nor is there any *static* view. Though omitted here for the sake of brevity, the temporal structure of the first-person must be understood. Subjectivity is a temporal gestalt. Though both demonstrative reference and holistic scheme are often easiest to think about in primarily spatial terms, they will be given a dynamic sense. The temporal structure of subjectivity has been the topic of many important philosophical studies, including Husserl's detailed analysis of internal time consciousness and Derrida's subsequent deconstruction of it. See Chapter 7.
23. For a powerful expression of the value of "situationism" for philosophy of mind, see J.T. Ishmael's *The Situated Self* (2007). For more on the perceptual-motor "enaction" of mind, see Noe (2004). I engage enactivism critically in Chapters 7 & 8. Finally, and given all of the above, it might be argued that the subject of an experience is an *unarticulated constituent* of that experience (Perry, 1986; Recanati, 2006). That is, experience is always "relativized" to a particular subject position, even though that position is not explicitly represented in the experience. But as it is found (and vigorously contested) in the literature, the notion of an unarticulated constituent is intended for the analysis of indexical propositions, not first-person experience. So I content

myself with the terminology of P* predicates and identification-free self-reference.
24. This response has been subsequently pursued by Cassam (1998) and also McDowell (1994).
25. This response has been adopted in a current book by Dan Zahavi (2014).
26. Actually it is a loosely unified and quasi-coherent set of partially overlapping positions, not a single subject position. But leave that to one side for now.
27. I recognize that this way of speaking does not do justice to the way the body figures as the self in experience. But I would argue that even visual perception of one's own body, along with proprioception and the bodily frame of reference, can be understood as a *specification* of subjectivity rather than as the immediate presence of the subject.
28. Ismael (2007, p. 70) writes, "...the fixed environment for thought for the human is the human body."
29. See Thompson (2007) for a statement of this transcendental version of the objection.
30. Teichert (2004), pp. 188, 177.
31. Atkins (2004), pp. 343, 348, 342, 354.
32. Atkins (2004), p. 364. Atkins stipulates that she uses the phrase "embodied consciousness" the way Merleau-Ponty did, to denote a conscious body-subject.
33. Teichert (2004), p. 183.
34. The label *strong narrativism* applies to more than one version of narrative theory. The key feature of the strong narrativism considered here is the conception of the narrative self as subject. A more defensible variety of strong narrativism refrains from applying the model to subjectivity, limiting its scope to the self. Rovane (1993) advocates this form of the view. See below.
35. Teichert (2004), pp. 178, 190.
36. Williams (1998), pp. 642.
37. Atkins (2004), p. 349. On this point Atkins and Teichert both follow Ricoeur, (1983–1988). However, please note that I do not attribute their version of the strong narrative subject model to Ricoeur. Nor does my argument stand or fall on any particular interpretation of his work. The model may be assessed independently without taking up the further question of whether Ricoeur endorses it. A full and fair interpretation of Ricoeur's monumental three-volume work *Time and Narrative* is beyond the scope of this essay.
38. Noggle (1998), p. 804.
39. Teichert (2004), p. 183.
40. Atkins (2000), p. 344.
41. Advocates of narrative such as Atkins (2004) and Teichert (2004) rely heavily on a distinction, found in Ricoeur (1983–1988), between two concepts of identity: 1. *idem* identity and 2. *ipse* identity. *Idem* identity is personal identity considered as *sameness* through time. Accounts of *idem* sameness aim to provide criteria for determining whether person x at time t_1 is the same as person y at t_2. This has been the focus of the literature on personal identity since Locke. Advocates of strong narrativism hold that *idem* sameness is insufficient by itself and that it must be underwritten by *ipse* selfhood. I move directly to consider this latter idea, which is the basis of the narrative subject model.

42. Teichert (2004), p. 185. cf. Ricoeur (1988), p. 246.
43. Atkins (2004), p. 348, Teichert (2004), p. 185.
44. Atkins, (2004), p. 343.
45. By contrast, Dennett (1992) holds an opposing, deflationary narrative theory in which "the self" is an irreal center of narrative gravity. His work is not on view here.
46. Teichert (2004), p. 188.
47. Cf. Kassin, SM. and Kiechel, KL. (1996), Horselenberg, R., Merckelback, H., and Josephs, S. (2003). See also Miller, et.al. (forthcoming), "Deconstructing self-blame following sexual assault: the critical role of cognitive processing."
48. Except when the term "I" is merely *mentioned* rather than used in the narrative. For example, "Then J.R. said 'I am in pain' as he fell to the floor." My argument here need only concern narrative *uses* of "I".
49. Philosophers (such as Pryor, 1999, and others) have made similar arguments about the epistemology of q-memory. Neo-Lockean accounts of personal identity based in psychological continuity face difficult counterexamples sometimes called "branch-line cases." It is debatable whether quasi-memories from another branch of the stream of consciousness are immune to error through misidentification. For my purposes in this essay, that debate can be safely put to one side.
50. Rovane (1993), p. 86.
51. Christofidou (1995), p. 223.
52. Rovane (1993), p. 93.
53. (1995, p. 229). In a related context, Campbell (1999) has argued that we cannot interpret our *own* subject-uses of the first-person pronoun by means of a "thinker of this thought" kind of description (pp?). This is because doing so would require knowing *who* the thinker is, thinking this thought.
54. Rovane (1993), p. 96.
55. Rovane (1993), p. 95.
56. Teichert (2004), p. 183. Brackets added.
57. Atkins (2004), p. 350.
58. An alternative interpretation of the narrativist claims just cited is that the narrative self is constructed on the *basis* of the first-person experience of acting, rather than that narrative selfhood is the necessary condition for agency and the experience of agency. But then the view is no longer strong narrativism. It collapses back into the weak variety, which provides no insight to the nature of subjectivity.
59. Perry (2002), p. 200. Note the connection between Perry's "being-at" and other, more ontological ideas such as "for-me" and "being-there."
60. Perry (2002), p. 208.
61. For present purposes, *file* can be read as "narrative self-understanding."
62. Nor does it carry any interesting implications for an account of personal identity. Having "first-person" experience does not require that the subject be a person. Many non-persons (e.g., animals) presumably have subjective experience.
63. This was a guiding insight of Merleau-Ponty's theory of embodied subjectivity. Yet Atkins interprets embodiment in terms of narrative, apparently reading Merleau-Ponty's phenomenological use of the term "*ipseity*" as equivalent to Ricoeur's hermeneutic use of that term.

64. Schechtman (1996, pp. 2–3). Schechtman's version of strong narrativism differs slightly from those of Atkins and Teichert in that she holds *character* to be narratively self-constituted rather than subjectivity per se. However, at certain points she comes very close to advocating the subject model, as when she speaks of temporal consciousness being constituted by narrative (pp. 137–45). Schechtman also shares with the narrative subject theorists the mistaken claim that narrative is not an identifying form of self-reference (p. 113). Here I consider only her claim that narrative is necessary to ground concern for the future.
65. Me* should not be confused with P* predicates. The * in "P*" marks the distinctively first-person perspective, while here the * in "me*" marks the self as a representational content or referent.
66. Strawson also argues for the possibility of a highly *ethical* life not narrated. I am sympathetic to this further point, but need not consider it here.
67. Sartre (1956), Campbell (2004), p. 475.
68. Thanks to an anonymous referee for this crystal clear way of putting it. Note also that this identifying role can be accounted for with a weaker form of narrative theory in which humans tend to tell autobiographical stories. This perfectly plausible version of narrativism may be sufficient for moral psychology, but not for a model of subjective consciousness.

3 Unconscious Subjectivity

1. This possibility raised by Zahavi (1999), and others.
2. A great part of Kriegel's book is dedicated to combating higher-order theories of consciousness, such as those offered by Rosenthal (2000) and Lycan (1997). Kriegel quite rightly objects that these are not really theories of subjectivity at all. But what they are is, they are theories of *awareness*! The right response, then, is not to kick against HOT/HOP theories because they do not account for subjectivity, but rather to *supplement* them with another account of subjectivity that is decoupled from awareness.
3. In fact there are many contemporary philosophers advocating the existence of phenomena related to unconscious subjectivity. Examples include: unconscious qualia (Burge 1997), experience of which we are unaware (Siewert 1998, Dretske 1993, Armstrong 1980), unconscious thought (Rosenthal 2000), unattended intermediate level representations (Prinz), and phenomenal-only-consciousness (Block 1995). Here I focus only on the last of these. I also give shout-outs to Prinz, Siewert, and others later in the chapter.
4. Obviously this is not the only way to distinguish among kinds of consciousness. Related distinctions among ways of being conscious or between different senses of "consciousness" can be found in, e.g., Levine (2001); Kriegel, (2006a); Rosenthal (2000); Baars (1997).
5. Thanks to Bill Robinson for the term "p-only-consciousness."
6. Block (1995), p. 231. Block clarifies that this should be relative to the capacities of the type of animal in question. He also says that even poor reasoning is sufficient (p.277).
7. (Graham & Neisser, 2000; Koriat, 2000).
8. This precise formulation is due to Kriegel (2009).

9. It seems that there can be p-only states that have no occurrent level of access whatever – at least that is the spirit of Block's work. Intuitions clash on this point. For interesting arguments in favor of the idea of "experience" that we are unaware of, see, e.g., Siewert or Dretske. I must admit, though, that it is difficult to maintain that unconscious subjective states have *zero* impact on consciousness. So, examples such as background stress or white noise are not perfectly decisive against the notion of peripheral awareness in Kriegel's sense. But whatever impact there may be, I maintain, remains implicit.
10. Kriegel (2006a, pp17) Kriegel (2009), Lycan (1996). Note that Levine (2001) is clearly committed to a strong version of the neo-Lockean conception too, since he holds that we *cannot fail* to be aware of qualia.
11. Could there be unconscious subjectivity in creatures with no consciousness whatever? Could there be zombie-subjects? Or perhaps it is only that conscious subjects can also have unconscious subjectivity? In the latter alternative, for every first-person perspective there would be consciousness, but not every mental state *for* that perspective would necessarily be conscious. I needn't choose between these alternatives at present. But zombie-subjects or something like them seem perfectly intelligible to me, and even a fairly plausible historical conjecture. There may be *dino-zombies* (we might call them "salamanders"). These would be phylogenetically primitive subjects with no metacognitive access to their own subjectivity. Very roughly, this *baseline subjectivity* would be enacted by the mesolimbic brain. See Chapter 6 & 7. See also Koch (2006).
12. For now, I use "experience" as a generic term that is neutral between the various possibilities of conscious awareness, p-only consciousness, unreportable experience, and unconscious subjectivity. I reserve the term "consciousness" for states of explicit awareness. This is a somewhat unhappy use of "experience" but I see no simple alternative, since in the literature p-only states are, unfortunately, referred to as conscious. But with the proper caveats in place it is not entirely misleading to think of unconscious subjectivity as a kind of "experience," i.e., as for-me.
13. Baars (1997, 2002).
14. Hampton reference about how to get animals to report on their introspective judgments.
15. i.e. Type Two errors. The term "misses" comes from J. David Smith, 2012.
16. In local re-entrance, there is isolated but relatively well-integrated neural activity, sustained by feedback ("re-entrant") connections in a neural network. This contrasts with global re-entrance, in which integrated neural activity has greater "online" depth and duration, and is associated more specifically with activity across the pfc and parietal lobe.
17. P.S. Churchland, *Neurophilosophy*. See also Flanagan (1992) on the *Natural Method*.
18. This phenomenon was first dubbed "selective attention" by Ulric Neisser and his lab in 1975.
19. Blockians can always return that all this merely adds to the case that what we *see* (are conscious of) far outstrips what we can attend to or report.
20. Part three of this chapter contains a full discussion of the cognitive unconscious and its contrast with the psychoanalytic model.

21. This characterization is drawn from Lakoff & Johnson (2000), Jackendoff (1987).
22. For more on the causal-mechanical nature of neurocognitive explanation, see chapter Eight, "Neural correlates reconsidered."
23. A parallel formulation in the phenomenological tradition is that there is a domain of conscious but non-reflective self-awareness, which can be uncovered by the phenomenologist. I avoid this formula, in part because it fails to do justice to the discovery of the unconscious (both cognitive & psychoanalytic). It is also too close to the account of subjectivity as peripheral awareness.
24. Marabou, (2012), p.42). This is not merely pre-reflective, but actually a non-*reflexive* level of self-reference.
25. At one point LeDoux refers to the self as "the totality of the living organism" (p.26). But this directly conflicts with the rest of his book, the primary thesis of which is fairly expressed in the title: *The Synaptic Self*.

4 What Subjectivity Is Not

1. This is a very incomplete list, naming only a few of the recent and salient contributions. Over the last century and more, the structure of the first-person has also been given extensive analysis by philosophers on both sides of the Atlantic including Husserl, Wittgenstein, Anscombe, Heidegger, Sartre, Merleau-Ponty, Evans, and Santayana.
2. With respect to the problem about qualia (the qualitative content of experience), one of the first two options may yet turn out to be the best.
3. Not all representationalists construe representations as information functions, and representational theories of mind need not be naturalist or materialist by any means. But in contemporary cognitive science the information-theoretic approach to representation is the default position and it is the approach for which consciousness, qualia, and subjectivity present explanatory problems.
4. In addition to Levine's text, the discussion in this paragraph and the next is partly drawn from a helpful conference session with qualophile and Levine collaborator Kelly Trogden of Virginia Tech University.
5. Though often downplayed or disavowed, this view of subjectivity – complete with epistemic baggage – is in fact shared by many contemporary philosophers, from Levine (2007) and Kriegel (2009) to Shoemaker (1997), Moran (2001), and Zahavi (2005).
6. The epistemology of subjectivity is also heavily emphasized in the literature on the so-called Knowledge Argument (Jackson, 1982).
7. The awareness requirement for qualia cannot be satisfied by just any conscious representation. That is, it is not sufficient that we be aware of something or other. The intrinsic qualitative content is said to be *determinate*, and this is precisely what renders inner awareness paradoxical. So not just any arbitrary awareness will do, on pain of making the awareness relation an extrinsic and indeterminate one. More below, and in section four, including a discussion of the self-representational view of subjectivity (Kriegel, 2009).

8. At this point in a prior version of this chapter I introduced what I called the *Cartesian Exception* to the representational theory of mind. The Cartesian Exception is that even though we might be wrong about how things are, we *cannot be wrong about how things seem*. That is, if I am wrong about reality, at least I am "right" about the appearances. In this way, Descartes held that ideas cannot be "materially false" (Descartes 2008 [1641])). With this, the epistemological trap of subjective duality is already laid for materialists. The truth-bearing content of the idea is treated as partly constitutive of the idea itself (the vehicle), as part of its "material." Below I pick up this historical thread with Berkeley and then Moore.
9. G.E. Moore (1903) believed that he had shown that at least there is no reason to think that this idealist slogan is true (p.435). Even this is probably claiming too much, but Moore's work does contain some telling critical insights, on which I draw presently.
10. Moore attributes this formulation to Taylor (1902).
11. This is a significant point of at which Levine parts company from Strawson (2006), Chalmers (1996), and others who offer panpsychist property dualism. On the latter views, although the existence of phenomenal properties does entail the existence of some being that has them, this certainly does *not* mean that these qualia must be known or perceived in any sense like that required by Levine's formula about subjectivity.
12. At this point in the exposition it is usual to introduce a distinction between transitive and intransitive consciousness, such as the one found in Rosenthal (2000) and elsewhere. However, because this section specifically concerns the idealist lineage of Levine's paradox, I will work with Moore's terminology. Moore's language will also connect more directly with the discussion of the Transparency Thesis in section three, below.
13. Some philosophers therefore advocate a *self-representational* account of subjective consciousness, in which states are subjectively conscious when they represent themselves in the right way (e.g., Kriegel, 2009. See also Levine, 2007). In the present context it is possible to regard self-representationalism as a variant of the well-known *phenomenal-concepts strategy* for explaining knowledge of qualia. See section four, below.
14. In general I do not object to the use of philosophical intuition as a method. In fact I accept, on intuitive grounds, the (different) view that there really are phenomenal properties. This view can only be justified on intuitive grounds. But I question whether the immediacy intuition is decisive ground for holding to the doctrine of Russellian Acquaintance. When it comes to our grasp of qualitative character I remain unconvinced because this further step requires an inference from phenomenological immediacy to epistemological immediacy, which does not follow. Also note that the immediacy intuition is the reverse of the equally misleading "transparency intuition" that guides the Transparency Thesis discussed in section three below. Cf Prinz (2012), p.14.
15. One reviewer objects that this is needlessly strong, asking, "Can we *never* directly know or grasp the contents of our experience?" In short, the answer is no, we never can. The idea of immediate subjective knowledge is what sustains the paradox. But in chapter two I argued that there is another, legitimate sense in which subjective thought should be understood as "direct" in a

way that distinguishes it from other forms of representation. The first-person perspective is immune to error through misidentification. But this is a subtly different matter. The sort of immunity displayed by the first-person emphatically does not accrue to the *content* or object of first-person judgment (in this case, to the judgment about the nature of the quale), but only to the prior issue of whether "I" am the subject making the judgment. See Campbell (2002, 1999), Pryor (1999), Moran, (2001, 1999), Shoemaker (1997), Evans (1982). Also see the end of section four, below.
16. If introspection is modeled as a self-scanning mechanism that generates a metacognitive representation of one's own mental states, then this representation can certainly be mistaken. For more on the idea that introspective awareness is a form of metacognition, high-order thought (HOT), or high-order perception (HOP), see e.g., Rosenthal (2000), Koriat (2000), Lycan (1997), Armstrong (1980). See footnotes 15, 20.
17. See also Rorty (1974).
18. See again Block (2007, 1996). For the record: Block's notion of *phenomenal-only consciousness* is a great idea, but it ultimately suffers from a variant of the same vacuity problem that afflicts the inner awareness approach to subjectivity. Just as it is empty to insist that subjects must be "aware" of all their qualia but needn't be held to any epistemic standard for "awareness," so too it is pointless to call phenomenal-only mental states "conscious" states. The trick, instead, is to distinguish for-me subjectivity from awareness/consciousness. See chapter three.
19. That is, the seemingly transparent nature of consciousness makes it appear as though blueness itself is *not* a mental fact. Hellie (2007) provides a nice historical exegesis of Moore's argument, showing that Moore did not in fact hold the Transparency Thesis but rather something close to the opposite. In short, Moore held that although consciousness *seems* transparent it isn't, and that this is a finding which can be corroborated by careful phenomenology or introspection. The object/property distinction is the key to Moore's larger argument, where it is employed to *reject* the tempting but false Transparency Thesis.
20. The phrase coined by Douglas Adams (1995 [1982]).
21. This is roughly the same move attempted by U.T. Place in his seminal piece on the central state identity theory, "Is consciousness a brain process?" (1956). According to Place, there is a "Phenomenological Fallacy" in supposing phenomenal properties to be properties of consciousness. In context, Place is clearly attempting to relieve identity theory of the burden of phenomenal qualities.
22. The better representational theory seems to be the one in which phenomenal properties track mind-independent properties but do not instantiate them. In order to function as a representation, of course, the stand-in need not intrinsically resemble the external property. Analogously, smoke neither intrinsically represents nor resembles fire, but it does contain information about fire. And this is also true of maps. Although some of the relational features of map representations structurally resemble the terrain mapped, the intrinsic properties of the map and the icons do not. For example, "you are here" red dots do not qualitatively resemble anything in the terrain, even while certain spatial relations with other mapped locations are preserved.

Similarly, phenomenal properties may be incidental intrinsic features of our cognitive maps of the external world. They preserve some relations between represented elements, much as any map preserves structural relations in the domain represented (P.M. Churchland, 2012, 2007).
23. A subsidiary negative thesis, just argued, is that subjectivity is also not the transparent awareness of external phenomenal properties.
24. Many versions of the phenomenal concept strategy have been offered, including well-known contributions by Papineau (2002, 1993), Tye (2000, 1995), Sturgeon (1994), and Loar (1997). The gloss presented here abstracts away from particular details, though it is most directly inspired by Tye (2009).
25. Thus, the phenomenal concept strategy has been extensively discussed in the literature about Mary the color-deprived neuroscientist, featured in the Knowledge Argument (Jackson, 1982).
26. And vice-versa: If I master a public concept without having had the corresponding phenomenal experience, I will be able to employ the deferential concept but not the phenomenal concept because I will have no token of the phenomenal property that constitutes it.
27. Note that on this point Kriegel's Self-Representational view is stronger than most formulations of the phenomenal concepts strategy and more faithful to Levine's paradoxical analysis. For Kriegel as for Levine, subjective inner awareness just *happens*; it is intrinsic to for-me mental states per se. The requirement that the subjective state be intrinsically self-representing means that the phenomenal property is constitutive of the awareness. I argued earlier that if self-awareness is intrinsic to qualia this entails that the awareness *cannot* misfire or misrepresent because the very existence of the phenomenal property suffices for the awareness. Unsurprisingly, Kriegel denies that his view has this epistemic consequence. His argument that subjective self-representations *can* go awry runs on the following example. "This very sentence is written in New Times Roman" is a self-representing sentence that can certainly be false, if the font is not New Times Roman (2005, p.39). This example is taken to show that there is nothing epistemically mysterious about the notion of self-representation. The problem with this example (and all examples of false self-representation) is that the representation relation in view is not an *intrinsic* relation of the sort required by the Self-Representational theory of subjectivity. The sentence does not represent itself *in virtue* of its font; the font does not constitute or establish the representational relation. The sentence's token-reflexive interpretation is governed by the phrase "this very sentence," an interpretation which is certainly not intrinsic to that phrase. So, even if the font in which a sentence is written can be called intrinsic to (constitutive of) the marks on the page, it is certainly *not* intrinsic to the content of the representation. Further, since representation in general fails to be an intrinsic relation, there can be no non-paradoxical (naturalized) theory of intrinsic self-representation.
28. While still defending the Transparency Thesis, Tye (2009) now holds that there are no special phenomenal concepts. One interesting reason he offers is that the phenomenal concepts strategy entails that even those who lack the requisite experiences are infallible about the content of those experiences! This is because they can't even formulate the relevant propositions and so

cannot be wrong in asserting them (Tye, 2009, pp.67–8). The phenomenal concepts strategy entails that a zombie who says, in the presence of a blue tapestry, "I see blue," does not say something false because, by hypothesis, the concept "blue" employed in the zombie's statement is a deferential concept that is used correctly. Tye's point here supports the Sellarsian argument I have just rehearsed. It adds to the case that the phenomenal concepts strategy cannot deliver physicalism about the requested kind of inner awareness.
29. Thanks to Evan Fales (University of Iowa) for forcefully expressing this concern during a lively conference discussion.
30. See (Matthews, 1977), for a nice historical treatment of this topic. See also footnote 9.

5 Introduction: Subjectivity in the Neurobiological Image

1. Of course, it is also possible to argue that these animals are persons after all, precisely on the basis of their subjectivity. But it is more plausible to conclude, on this basis, that persons are not the only creatures worthy of moral consideration.
2. It is clear that Sellars understood the scientific image to be mechanistic. For example, he wrote of the "iffy" or counterfactually supporting and dispositional behavior of physical entities, and how these iffy behaviors are explained in the scientific image by sufficiently specific underlying systems of postulated entities, connected in the right way to overtly observable phenomena (p.24).
3. It must be admitted that the mechanistic explanation does appeal to affect, which ultimately has a qualitative component that remains unexplained. But given that there *are* affects, the neurobiological image explains how they function in habitual activity.

6 The Science of Subjectivity

1. Those few writers who have argued otherwise, such as Jaak Panksepp, have also lacked a clear distinction between conscious awareness and subjectivity. But, as discussed below, Panksepp (2012) has recently seen the importance of distinguishing "awareness" from his brand of affective consciousness.
2. In addition, it has been argued that this perspectival structure has a *stabilizing* function. It is an enduring anchor or 'center' for perceptual and affective experience. Thus, the "I" of the first-person perspective occupies the point of "maximal invariance" in an integrated conscious perspective. Metzinger (2000), p.297. This seems basically correct, except that subjectivity isn't nearly as "centered" or unified as tradition has it. Instead, there is a simple collection (aggregate) of overlapping and interacting systems of orientation and engagement, all of which just happen to be located within a single animal.
3. This is perhaps the most fundamental finding of the phenomenological tradition, expressed by Heidegger as *sorge*. Despite many other mistakes,

Heidegger correctly identified this concernful involvement as the core of mental life and distinguished it from consciousness per se.
4. Do not mistake this for a claim of mind/brain identity. But the brain (CNS) is the deepest layer of the embodied onion; the more you peel away, the stinkier it gets.
5. This process is certainly *aesthetic* but by no means *transcendental*.
6. For further discussion, see chapter seven on putting the *neuro* in neurophenomenology and chapter eight on the NCC reconsidered.
7. Or, conversely, spatiotemporally articulated affection.
8. Subjectivity (in this baseline form which we share with animals) is not a recent, sophisticated modern malady invented by European priests and scholars. Instead, the "mummeries with which we are upbraided" (Hume) in our modern and postmodern condition are no more than a few historically particular *ways* of coping, which we learn by mapping with affective valence. They could change completely and there would yet be baseline subjectivity in a recognizable first-person form. And there would still be first-person subjectivity even if the category *person* were someday completely deconstructed and done away with.
9. Experiments in olfactory perception often utilize only static odorants, not ecologically valid changing chemical arrays. According to Jacobs, that sort of research is asking the wrong questions, and cannot test the OS hypothesis. See, e.g., Murthy, VN (2011).
10. "Far better an approximate solution to the right question...than an exact answer to the wrong question..." An evo-devo rallying cry – attributable to John Tukey.
11. Her analysis also takes account of whether the animal later evolved non-olfactory methods of navigation such as vision or echolocation.
12. The former element is associated with Tolman, while the latter was contributed by Gustav Kramer.
13. Again, the PMT is but one proposed mechanism to explain how animals navigate. Others are possible, and for my purposes nothing hinges on the particulars of this theory. For example, it is arguable that the third, integrated map is unnecessary because "short cuts" may be discoverable more directly from gradient and bearing information, through a process of vector subtraction. Or it may turn out that there are many more than three such maps. I need not evaluate these competing possibilities here. The point is to give the reader a feel for research on navigation in neuropsychology within the evo-devo framework.
14. In fact, the primary function of the hippocampus was once thought to be olfaction. Sarnet & Netsky (1981).
15. For a more complete discussion of dynamic systems theory, see the next chapter.
16. Freeman is at least right in one respect: the system's *overall* state space is not represented in the actual mechanism, as in an "internal world-model." Instead, the head maintains an orientation with respect to the environment by finding stability among all the parameters in the dynamic interaction. Note that state "space" itself is a quasi-visual metaphor used by researchers to think about the range of parameters in a set of equations. It is a graphic representation of the solution set for all possible functions in the model.

The points in this virtual state space represent possible states of the brain, but the brain does not represent *them* (at this level). It just goes through the stabilizing process...ambient conditions, including my embodied architecture are *unarticulated constituents* of my subjective point of view (cf Perry 1993). That is, my first-person experience *concerns* the animal/environment relation without explicitly representing either the animal or environment as a whole. "If all that an agent needs is to know what is going on in her neighborhood, it suffices to give her a source of information whose states have an implicitly indexical spatial content." (Ishmael, 2007, p.18) Thus, for situated cognition, it is not so much what's inside the head as what the head's inside of (Blackburn). For humans, the possibilities for action are enhanced beyond those of most other animals by our specialized abilities in imagination and expectation, which are elaborated in neocortex. We are capable of explicitly representing the environmental layout in thought, and of *orienting ourselves to the internal model*. In this way we free ourselves of immediate context (Ismael, 2007, pp63). But the basic activity of space-time orientation that stands at the root of subjectivity is undertaken by all vertebrates, and *at that level* the world can serve as its own model.

17. Freeman, 2001, p.103. Neither of these researchers can quite account for *temporality* in the first-person perspective. See the full discussion in chapter seven, putting the *neuro* in neurophenomenology.
18. Notice that Freeman cites Jacobs' PMT as an ally here, but I find nothing in Jacobs' writing to indicate that she considers cognitive mapping to be merely metaphorical. Instead, she consistently argues that maps in the olfactory-limbic system are the mechanisms of animal navigation.
19. This basic notion goes by countless names in Panksepp's writing, all of them equally unsatisfying. I have abstracted the underlying idea and dubbed it with this name.
20. Panksepp states: "...localized electrical stimulation of the brain (ESB)...provoke[s] similar types of instinctual behaviour patterns, and accompanying feelings, in all animals across all mammalian species that have been studied. Such stimulations in the neocortex only impair functions, and generate no clear affects. (Panksepp (2012), pp.11–12.]
21. Ikemoto (2010a), Brain reward circuitry beyond the mesolimbic dopamine system: a neurobiological theory, *Neuroscience and Behavioral Review* **35** (2), pp.129–150. Panksepp (1998), pp.29, 189, 239, 243.
22. Historically, dual-aspect monism is the name for Spinoza's ontology in which God is Nature, the world is a divine mind, and we conscious creatures are its eyes and ears. The somewhat moderated descendent of this ontology in contemporary philosophy of mind is usually called *property dualism*. Though property dualism might well be true, it is incompatible with physicalism, and therefore seems untenable to most empirically-minded researchers.
23. Panksepp (2012), p.12. See also Gallagher (2008), p.91, and elsewhere. There will be much more in the next chapter on homology and its distinction from analogy. For the moment, pay no mind to the buzzing produced by the use of both "analogy" and "homology" in this paragraph.
24. Farah (2010). For the record, I reject Farah's claim that Mill's Argument from Analogy is strengthened by the addition of a metaphysic of supervenience, because the supervenience relation itself remains opaque.

25. Recently, Panksepp (2012) himself has seen the value in distinguishing his proposed form of affective consciousness from awareness. He now recognizes that the SEEKING system and the other affective modes are not – and need not be – sufficient for consciousness in its full-fledged metacognitive, socially and linguistically mediated form. This recognition is reflected in a new emphasis on the phrase "primary emotional-affective processing systems," from which I derive the phrase "primary affective modes."
26. Freud's term "preconscious" might be better here.

7 Putting the *Neuro* in Neurophenomenology

1. See also, Levine (2007) In addition, Jackson (1994) argues that materialist accounts of mind need to provide some good old fashioned conceptual analysis in order to "find the mind" in the material world. I agree completely, and this is what was attempted in Part One of the book. No doubt my analysis of baseline subjectivity somewhat alters received ideas about subjectivity. And this dialectical or coevolutionary process of "relocation" is what neurophilosophy is all about; it is necessary for progress in philosophy of mind. See the Postscript on "neurophilosophy, naturalism, and subjectivity."
2. Thus, whether Mary-before would know what-it-is-like to see color depends on just how omniscient and ideal one imagines her to be. If Mary is also allowed to know the complete evolutionary and developmental history of an individual animal up to the moment it has a particular experience, then Jackson's Constraint implies that she ought to be able to predict the changes in the content of that animal's first-person perspective from physical changes alone.. I hold that this is conceptually possible. But note that even this would not *produce* the animal's experiential perspective in Mary. Her brain will not *thereby* enact the relevant subjectivity.
3. To clarify, I do not accept a priori physicalism as a general approach to scientific explanation. Briefly, a priori physicalism is the view that explanation is a modal relation, a matter of the derivation of the explanandum from the laws of physics. Instead, I take a broadly mechanistic interpretation of the *semantic view of theories* (see below, and chapter eight). But here I adopt Jackson's Constraint for the sake of argument, in order to make a constructive connection with the literature on the explanatory gap.
4. Jackson, (2003), pp.6. I give the example as Jackson presents it. But as it stands the example is woefully incomplete because it *omits* the crucial functional analysis of temperature that is supposed to provide the connection to the physical science. Without this, it is implausible that any such derivation has ever convinced anyone of the physics of temperature. Of course the "so and so" phrases in Jackson's shorthand might be filled in several ways. Temperature might be functionally defined as that which satisfies T in the ideal gas law: $PV = nRT$. But this is cheating a little since n refers to the number of molecules and thus arguably makes indirect use of the notion of mean kinetic energy. This difficulty in connecting *purely* functional analyses to contemporary empirical findings indicates that "a priori physicalism" is not so a priori after all, and not a promising philosophy of science.

5. Here it is also worth recalling that U.T. Place once suggested that "we can identify consciousness with a given pattern of brain activity, *if we can explain* the subject's [experiences] by reference to the brain processes with which they are correlated" (Place, 1956, p.44, emphasis added). The empirical account suggested by the evo-devo framework is not a version of central state identity theory (!), but arguably it *is* a kind of whole-organism identity theory, and this is what distinguishes it from straight functionalism and qualifies it as an "embodied" approach. See below on "modular embodiment." See also chapter eight for a causal interpretation of the NCC, and the relation between causal mechanisms and phenomenology.
6. Ehreshefsky (2009), Love (2007), Griffiths (2007), Amundsun & Lauder (1994). But see Neander (2002) for an opposing view. There is also a further debate in the literature about the homology concept, reminiscent of an earlier debate surrounding the species concept. Some hold that homology must be historical by definition, while some argue that historical explanations of homology are distinct from the observable phenomena per se (Griffiths, 2007). There is also debate about whether the essence of the homology concept is taxic, phyletic, transformational or developmental. I basically adopt the view of homology found in Ehreshefsky (2009) and in Love (2007).
7. Brigandt & Griffiths (2007) compare the character/state distinction to the determinable/determinate distinction.
8. Homology can also exist independently at various *levels*, genetic, developmental, morphological, and behavioral. A higher-level homologue (behavioral or morphological) can be produced by genetic or developmental processes that are not homologous. Ereshefsky (2012, 2007) argues that this "hierachical disconnect" can be a key to explaining multiple realization in neuropsychology.
9. To clarify, the two kinds of thinking are compatible.
10. This objection, of course, echoes the traditional functionalist line in philosophy of mind, according to which the actual implementation story can add nothing to the conceptual understanding of mental states. And it is precisely this traditional version of functionalism for which consciousness is the insuperable problem. See next note.
11. This is a biological version of the zombie arguments in philosophy of mind. There, the reasoning rests on modal intuition (Chalmers, 1996; Kripke, 1972). Here the argument runs on the distinction between homology and analogy. In both cases, the point is that the output function by itself does not exhaust all there is to explain about the mind.
12. Further, it is not necessarily "chauvinism" to explain something as a historical particular, especially when the case encompasses a swath of world history dating from the Cambrian explosion.
13. For a mathematical model of "temporality" construed as the formal function of temporal sequencing abstracted from any implementation particulars, see Grush (2006).
14. Developmental modules and homologues are closely related, and it is tempting to simply identify the two. Informally, I write as if the homologue is what is produced by the developmental module. This intuitive way of thinking becomes problematic because the modules themselves can be

homologous to one another. But nothing in my use of these ideas hangs on these niceties. See Wagner (1996).
15. The term "quai-independent" was coined by Lewontin (1970). For more on the idea of a developmental lock, see Wimsatt (1986).
16. Also note that much neural anatomy, such as the columnar organization of the neocortex, consists in *iterative homologues*. "Iterative homology" refers to the fact that homologous structures can recur within one and the same individual, recruiting the existing genetic and developmental pathways in repeated and varying fashion (sometimes these reiterated structures are called "homonomous"). A simple example of iterative homology is the vertebrate spine. The OS hypothesis suggests that limbic structures such as the hippocampus may be evolutionarily modified reiterations of the olfactory bulb (see chapter six).
17. I omit to discuss Hutto's approach in any detail,, but it seems poorly motivated. While it is certainly true that there are persistent conceptual questions about mental representation, rejecting the very idea creates far more problems than it solves. For other "radical" research on embodiment in Hutto's vein, see Wilson & Galonka (2013).
18. Cf Georgialis (2005). Also see Neisser (1997), *The Embodied Approach to Cognition: A Defense*. My Ph.D. dissertation.
19. Clark is consistently dismissive of phenomenological and/or introspective considerations.
20. And of course there are several versions of enactivism, too. I omit discussion of Alva Noe, for example.
21. Contrast this ontological "passage" with the epistemic passage from third to first-person perspective required by Jackson's Constraint.
22. This definition is deceptively simple, since the three-place relation it names can track any number of parameters.
23. I omit messy detail. For one thing, mathematical models are typically supplemented by "toy models" or simulations. Today these are almost always computer visualizations, but the globes, stick figures and solar system mobiles still found in classrooms are also examples.
24. For current examples see Craver (2007) or Bechtel (2008).
25. Also, even in non-self-organizing dynamic systems (with pre-existing manufactured parts), it is quite clear that the ascription of functional interactions among the parts needn't assume that system behavior is a straight linear sum of these components.
26. It is worth noting that the doctrine expressed by Maturana & Varela also runs counter to the vast weight of practice in experimental neuroscience.
27. The project of genetic phenomenology is to inquire into the auto-affective, non-egoic origins of subjectivity, including temporality and territoriality. This late-period genetic phenomenology is highly relevant to the project here and to any other deconstruction of transcendentalism. Although there is no place for a full discussion of it here, this genetic approach is one key to a better and truly naturalized phenomenology.
28. Gallagher & Zahavi (2012, p.85). The authors draw their gloss from Husserl (1962/1977, p.202). This canonical statement by Gallagher & Zahavi is exactly similar to the one relied upon in Neurophenomenology®.

29. The phrase "minimal line of tension" in this context is due to Nicolas de Warren.
30. James' term was *penumbra*, while several contemporary cognitive scientists speak of *working memory* (Baars, 2002).
31. Ishmael (2007), p.142. As Gallagher & Varela (2003) put it, temporality is shaped by boundary conditions fixed through action and engagement. In Ishmael's phrase, The permanent context of human thought is the human body.
32. See chapter two. The "specification" need not be particularly precise, coherent, or stable. But in the first-person, things tend to happen *as if* unified and well-specified. Even this should not be taken too strongly – experience often feels *as if* fuzzy and fragmented! These variations on the here and now are but different character states of the first-person perspective.
33. I do not suggest that Zahavi would endorse my larger view of subjectivity, or the project of explaining it. The point is just that, in Husserl's analysis, temporality is first-person or for-me in the relevant sense. But Zahavi also endorses a version of the "awareness conception" of subjectivity discussed in chapters three and four. For Zahavi, PRSC is always understood as a self awareness that is intransitive. But in my view, *awareness* is always transitive. Subjective phenomenality need not be transitive, however, and subjectivity takes place when there is something it is intransitively like for the animal. Conscious awareness takes place when the animal then becomes transitively aware of intransitive subjectivity.. In addition, Zahavi (2005 and elsewhere) distinguishes "reflective" from "reflexive" awareness, and holds that PRSC is still *reflexive* self-awareness. I disagree because the very notion of *self-awareness* is epistemically loaded in a way that requires reflection, not mere reflex.
34. Note that "what happens" – the *way* the microstructure of temporality performs its function (plays it's causal role) – is an *activity* in the sense specified above and by Love (2007). A similar causal role could be played by any number of other activities in other systems, and the same activity could play a different causal role in another system.
35. See also Gallagher & Zahavi (2012 [2008]).
36. Thompson characterizes neurophenomenology as an offshoot of enactivism, which is his name for a larger philosophy. Nothing hangs on it, but I prefer to reverse this priority, taking neurophenomenology as a general field and enactivism as a specific approach to it, namely, the one that generates Neurophenomenology®.
37. Husserlians object to the characterization of phenomenological reflection as a form of introspection (but see Bayne, 2004). I only use this comparison for expository purposes, to give the reader a flavor of the experimental practice.
38. Lutz, et al (2002), Van Gelder (1995), Rodriguez et al (1999). See also Lloyd (2012, 2002), Thompson, Lutz & Cosmelli (2005), Lutz & Thompson (2003), Gallagher & Varela (2003), Varela (1999). I omit to discuss this aspect of Neurophenomenology®. For critical notices, see Grush (2006) or Bane (2004).
39. In addition to the three elements just described, Neurophenomenology® is also committed to a *Method of Reciprocal Constraint* (MRC) in which first and third-person investigation are to be placed in conversation, each guiding and

revising the findings of the other (Thompson, 2007, p.340). MRC resembles the methods of Consilience (Wilson, 1998), Natural Method (Flanagan, 1992), Co-evolutionary Method (Churchland, 1986), and Reflective Equilibrium (Rawls, 1971). I limit myself to one comment on the MRC. If the method is to earn its keep, it must be genuinely *reciprocal*. The primary concern in the Neurophenomenology® literature has been to make sure phenomenological method has a respected voice in the conversation. But the test of the MRC is whether "reciprocal constraint" will ever run the *opposite* direction. Are there any phenomenological claims that might be revised or reconsidered in light of what we know about neuropsychology? Or not?
40. Smart credits this thought to Max Black as a question in colloquium. See also Block (2006).
41. In philosophy of science this general approach is known as the *semantic view of theories*. It is sometimes informally called "the model model" of explanation. See Suppes (1967), Giere (1999).
42. I gloss over distinctions among kinds of functionalism. Only some functionalists (so-called "realizer functionalists") identify the mental state with the neural state that plays the function. "Role functionalists" identify the mental state with the second-order relation – the functional role itself – and not the realizer that plays the role. Realizer functionalism can be understood as a version of the central state identity theory, but role functionalism cannot. But the qualitative contents of consciousness don't seem to have a function anyway, so neither sort of functionalism will avail in this context.
43. Cf. Thompson (2007), p.337. Grush (2006) is highly critical of this idea, and I (on somewhat different grounds) I think it displays a peculiar form of misplaced concreteness.
44. Note that this is where a committed Husserlian could part ways with enactivists such as Thompson, pursuing a *genetic phenomenology* and inquiring back into the precursors of embodied subjectivity.
45. For a similar point, see Grush (2006). For another example of research on neuroscience and temporality that runs into this kind of problem, Lloyd (2012).
46. See chapter six for an introduction to the ideas of SEEKING and the bearing map.
47. For additional reason to believe that such models can be *identification free* in the sense required, see, e.g., Ishmael (2007), Perry (2002), and Cassam (1997).

8 Neural Correlates of Consciousness Reconsidered

1. The phrase "neural correlates of consciousness" was coined by Crick (1990), but it is the definition of it in Chalmers (2000) that is under consideration here. For more contemporary examples of neurobiological research on content consciousness, see, e.g., Koch (2004) Singer, (1995), Reese et al (2002), Edelman, G., & G. Tononi (2000).
2. Significant complications arise when it is asked, under what conditions will the neural state suffice? Presumably, if neurons are excised from the brain, placed in a dish, and given electrochemical stimulation, their activity will

not suffice for content consciousness. Here Chalmers resorts to the ceteris paribus clause *under conditions C*. But defining the exact range of cases in which a putative NCC ought to be sufficient for the corresponding conscious state presents a thicket of difficulties. After dedicating a sizable chunk of his paper to discussing this issue, Chalmers tentatively concludes that *"... conditions C might be seen as conditions involving normal brain functioning, allowing unusual inputs and limited brain stimulation, but not lesions or other changes in architecture. Of course the precise nature of conditions C is still debatable.... But I think the conditions C proposed here are at least a reasonable first pass, pending further investigation."* (Chalmers, 2000, p. 31). In part, the difficulty arises from the functional and developmental plasticity of the cortex. Brains with lesions or other structural alterations may find unique ways to function and may have unique NCC profiles. Thus, data from lesion studies must be interpreted very carefully, since observations of a heavily damaged brain may not identify activation that is sufficient in a more typical neural system. Also note that seeking the specification of the range of conditions under which a neural state correlates with a particular conscious content is very much like asking for a specification of the conditions under which a data structure in an information system bears its content. The answer is that a data representation bears its content in *that* system and in other, relevantly similar systems. Specification of the relevance parameters for the interpretation of information has, of course, confounded research in cognitive science and artificial intelligence for decades (Dreyfus, 1992). One motivation for rethinking the correlate idea is that much contemporary neuroscience also demands a rethinking of the classical conception of representation that it takes for granted. The internal temporal dynamics of the brain rarely fit the description of well-specified discrete representational contents. Instead, dynamic models of neural activity suggest that we conceive of a trajectory through state space as "a temporally drawn-out pattern of multiple 'representations' being simultaneously partially active" (Spivey, 2007, p. 5. See chapter seven).

3. The three global features they discuss are closely related. Each is meant to show that conscious visual content appears as part of a larger structure of perception and action, as a particular figure against a larger background frame. Thus, visual consciousness is not simply a patchwork of independent atoms, each of which corresponds to some particular neural activation in visual cortex. This is a familiar phenomenological and gestaltist point, but none the worse for wear.

4. In his comments on Noe & Thompson's focus piece, Metzinger defends the neural correlate idea by denying that it is committed to the matching content assumption. I agree that the neurobiological research itself is not committed to the matching content assumption. But Chalmers' definition remains committed to it in virtue of the sufficiency requirement. Without the matching content assumption, the correlation relation obtains only between global subjective states and global states of the nervous system (Metzinger 2004, p. 71).

5. Chalmers' notion of a core NCC should not be confused either with Damasio's (1999) "core consciousness" or with Edelman & Tononi's (2000) "dynamic

core." Here the problem stems from the direct transposition of Shoemaker's idea of a core realizer for a belief state to phenomenology, and identifying the putative core experience with the representation of the preferred stimulus. But the holistic, gestalt nature of conscious content is not a problem that Shoemaker's approach needs to contend with.

6. Even with the addition of the criterion that the INUS condition must also be "effective" in this way, it remains conceivable that B is in reality the cause of A, not the reverse.
7. cf. Craver (2007), Cartwright (2007), Pearl (2000), Glymour (1994), Lewis (1973). See also van Fraassen (1980) for more on the notion of an *effectual difference*.
8. Craver doesn't agree with Mackie in every respect, of course. The present comparison is based on their shared ontic approach to causality and its implications for experimental design and control.
9. Craver (2007), p. 200. See also Godfrey-Smith (2003) Giere, (1999);Hacking (1983).
10. *Optogenetics* is another technique for direct neural manipulations, which can reach areas deep within the brain in a way that TMS cannot. In optogenetics, specific neural populations are first genetically modified to be sensitive to certain wavelengths of light, such that they can be made to selectively fire by shining light on the animal. Spectacular results have been obtained, "switching" e.g. aggressive behaviors on and off by switching the lights in the rat's cage. See, e.g., Anderson (2012).
11. Indeed, very many questions can be raised about the larger metaphysical framework for the philosophy of consciousness. What type of materialist ought one be, A, B, C, D, E or F (Chalmers, 2002)? What constitutes real being and real causation (Nagarjuna & Garfield, 1995)? What is the relevance of dual-aspect monism (Vimal, 2009)? While these are all important and interesting questions in their own right, they are precisely the sort of questions that can be – that *must* be – set to one side while the focus is shifted onto philosophy of science and away from the metaphysics of mind. Whether the metaphysician has any use for this data, whether or to what degree the neuroscientific assumptions about mechanistic explanation are accepted, these are further issues far beyond the scope of this paper and in fact beside the immediate point.
12. More might be said about the apparent tension between the idea that neural states constitute conscious states and the idea that they cause changes in conscious states. This is not the place to pursue this matter fully, because the present point does not much depend on it. That is, whether the neural mechanism in view is constitutive, causal, or both, does not affect the basic point that it is not a metaphysically neutral "correlate." For my part I suspect that this is a false dilemma between constituents and causes. What looks like a cause when considered from a linear perspective with a short time-frame will look like a constituent when treated within a dynamic systems perspective and longer time-frames.
13. Or at least, the brain should be seen as a "minimally decomposable" system. The practical difference between these two possibilities is unclear.

Postscript: Neurophilosophy, Darwinian Naturalism, and Subjectivity

1. There are many more, too, from neuroeconomics to neurocriticism and even a "brain based philosophy of life" (Gazzaniga, 2005).
2. And conversely, historicism as a method can be extended *into the hard sciences* from the humanities and social sciences.
3. Except in the limiting case that *ought implies can*. If the facts categorically prevent me from following some principle, then it cannot be the case that I ought to follow that principle. But even here, the burden is on the naturalist actually to prove the impossibility of acting according to the putative ideal, without merely adverting to the general claim of absolute determinism.
4. I omit to defend the fact/value distinction. But the careful naturalist can avoid the naturalistic fallacy even without adhering to any robust version of this distinction. While it is perfectly true that no facts are ever described from a perfectly value-neutral perspective (a point which is near and dear to the heart of the account of subjectivity offered in this book), the naturalist need only *acknowledge* this point as a problem in determining the normative import or critical assessment of any particular "factual" claim. Thus, the critique of implicit norms in pseudo-naturalistic writing may be useful for naturalists as much as for anti-naturalists.
5. Davies carefully skirts the genetic fallacy. As we have seen, to show that an idea descends from some questionable source does not, by itself, constitute a refutation. Accordingly, Davies does not conclude that concepts dubious by descent *must* be false. But he clearly thinks they are false, and he thinks this partly on the basis of their genealogy.
6. For a priori physicalists such as Jackson (1994) or Chalmers (2014), locating x means something stronger – deductively *deriving* x from physical law. See chapter seven.
7. Thompson (2007) explicitly connects autopoeisis to subjectivity and traces it back to Kant. Kant is one of the primary ancestors identified in Davies' genealogy of concepts dubious by descent.
8. He culls this characterization of subjectivity from Jackson's infamous Knowledge Argument (1982).

Works Cited

Adams, D. (1995 [1982]). *Life, the Universe, and Everything*, Del Rey Books, London
Alcaro, Huber, & Panksepp (2007). Behavioral functions of the mesolimbic dopaminergic system: An affective neuroethological perspective. *Brain Research Reviews* 56: 283–321
Alcoff, L. (1988). Cultural feminism versus post-structuralism: The identity crisis in feminist theory. *Signs*, 13 (3), 405–436
Anderson, D.J. (2012). Optogenetics, sex and violence in the brain: Implications for psychiatry. *Biol. Psychiatry*, 15; 71(12): 1081–1089
Anscombe, G.E.M. (1975). The first person. In Guttenplan, S. (ed.), *Mind and Language*. 45–65, Clarendon Press, Oxford
Armstrong, D. (1980). *The Nature of Mind and Other Essays*. Cornell University Press, Ithaca, NY
Atkins, K. (2000). Personal identity and the importance of one's own body: A response to Derek Parfit, *International Journal of Philosophical Studies* vol. 8 (3): 329–349
Atkins, K. (2004). Narrative identity, practical identity, and ethical subjectivity. *Continental Philosophy Review* 37: 341–366
Baars B.J. (2002). The conscious access hypothesis: Origins and recent evidence. *Trends in Cognitive Science* 6 (1): 47–52
Baars, B.J. (1997). Treating consciousness as an empirical variable: The contrastive analysis approach. Block, N. Flanagan, O. and Guzeldere G. (eds) *The Nature of Consciousness: Philosophical Controversies*. Bradford/MIT Press, Cambridge, MA
Bargmann (2006). Comparative chemosensation from receptors to ecology. *Nature* 444: 295–301
Bayne, T. (2004). Closing the Gap? Some Questions for Neurophenomenology. *Phenomenology and the Cognitive Sciences*, 3/4: 349–64
Bayne & Chalmers (2003). What is the Unity of Consciousness? in A. Cleeremans (ed.), *The Unity of Consciousness: Binding, Integration, Dissociation* Oxford
Bechtel, W. (2008). *Mental Mechanisms: Philosophical Perspectives on Cognitive Neuroscience*. Routledge Press, New York
Bechtel, W. & Abrahamsen, A. (2011). Complex biological mechanisms: cyclic, oscillatory, and autonomous. In Hooker, (ed.), *Philosophy of Complex Systems*. *Handbook of the Philosophy of Science, Volume 10*. 257–286, Elsevier, Oxford
Bechtel, W. & Abrahamsen, A. (2005). Explanation: a mechanist alternative. *Studies in History and Philosophy of Biology and Biomedical Sciences*, 36: 421–441
Bermudez, J.L. (1998). *The Paradox of Self-Consciousness*, MIT Press, Cambridge, MA
Berridge, K. (2012). From prediction error to incentive salience: mesolimbic computation of reward motivation. *European Journal of Neuroscience*, 35: 1124–1143
Bickle, J. (ed.) (2009). *The Oxford Handbook of Philosophy and Neuroscience*. Oxford University Press, New York

Bickle, J, Mandik, P, and Landreth, A, (2012), The Philosophy of Neuroscience, In Zalta, E.N. (ed.), *The Stanford Encyclopedia of Philosophy* (Summer 2012 Edition), URL = http://plato.stanford.edu/archives/sum2012/entries/neuroscience/

Block, N. (2011). Perceptual consciousness overflows cognitive access. *Trends in Cognitive Sciences* 15 (12): 567–575

Block, N. (2007a). Consciousness, accessibility and the mesh between psychology and neuroscience. *Behavioral and Brain Sciences* 30: 481–584

Block, N. (2007b). *Consciousness, Function, and Representation: Collected Papers Volume 1*, MIT Press, Cambridge, MA

Block N. (2006) Max Black's objection to mind/brain identity. In Zimmerman (ed.) *Oxford Studies in Metaphysics* v.2, 3–78. Oxford University Press, New York

Block, N. (1995). On a confusion about a function of consciousness. *The Behavioral and Brain Sciences*, 18 (2): 227–287

Bohning, D.E. (2000). Introduction and overview of TMS physics. In George & Bellmaker (eds), *Transcranial Magnetic Stimulation in Neuropsychiatry*. American Psychiatric Press, Washington, D.C.

Boothby, R. (2001). *Freud as Philosopher: Metapsychology after Lacan*. Routledge Press, New York and London

Boring, E.G. (1950, [1929]). *A History of Experimental Psychology*. Appleton-Century-Crofts, New York

Brandom, R. (2015). *From Empiricism to Expressivism*. Harvard University Press. Cambridge, MA

Brandon, R.N. (1999). The Units of Selection Revisited: The Modules of Selection. *Biology and Philosophy* 14: 167–180

Brigandt I. (2007). Typology now: homology and developmental constraints explain evolvability. *Biol Philos* 22: 709–725

Brigandt, I. & Griffiths, P.E. (2007). The Importance of Homology for Biology and Philosophy. *Biol Philos* 22: 633–641

Bruner, J. (1993). *Acts of Meaning*. Harvard University Press. Cambridge, MA

Burge, T. (1997). Two kinds of consciousness. In Block, N., Flanagan, O. and Guzeldere, G. (eds) *The nature of consciousness: Philosophical controversies*. Bradford/MIT Press, Cambridge, MA

Campbell, J. (2004). What is it to know what "I" refers to? *The Monist*, 87 (2): 206–218

Campbell, J. (2002). *Reference and consciousness*. Oxford University Press, New York

Campbell, J. (1999). Immunity to error through misidentification and the meaning of a referring term. *Philosophical Topics* 26 (1&2): 89–104

Carroll, S. (2005). *Endless forms most beautiful: The new science of Evo Devo*. Norton, New York

Carruthers, P. (2009). Invertebrate concepts confront the Generality Constraint (and win). In Lurz, R. (ed.), *The Philosophy of Animal Minds*, CUP

Carruthers, P. (2005). *Consciousness: Essays from a Higher-Order Perspective*. Oxford University Press, New York

Cartwright, N. (2007). *Hunting Causes and Using Them*. Cambridge University Press, New York

Cassam, Q. (1997). *Self and World*. Oxford University Press, New York

Chalmers, D. (2002). Consciousness and its place in nature. Reprinted in Chalmers, (ed.) *Philosophy of Mind: Classical and Contemporary Readings*. Oxford Press, New York

Chalmers, D. (2000). What is a neural correlate of consciousness? In Metzinger (ed.), *Neural Correlates of Consciousness: Empirical and Conceptual Questions*. MIT Press, Cambridge, MA
Chalmers D. (1996) *The conscious mind: In search of a fundamental theory*. Oxford University Press, New York
Charvet & Finlay (2012). Embracing covariation in brain evolution: Large brains, extended development, and flexible primate social systems. Hofman, M.A. and Falk, D. (eds) *Progress in Brain Research*, 195: 71–87
Churchland P.M. (2012). *Plato's camera: How the physical brain captures a landscape of abstract universals*. MIT Press, Cambridge, MA
Churchland, P.M. (2007). *Neurophilosophy at Work*. MIT Press, Cambridge, MA
Churchland, P.S. (2013). *Touching a Nerve*. Norton & Co., New York
Churchland, P.S. (1994). Can Neurobiology Teach Us Anything about Consciousness? *Proceedings and Addresses of the American Philosophical Association*, 67 (4): 23–40
Churchland, P.S. (1986). *Neurophilosophy: Toward a Unified Science of the Mindbrain*. MIT Press, Cambridge
Clark, A. (2013). Are we predictive engines? *Behavioral & Brain Sciences* 36: 181–253
Clark, A. (2010). *Supersizing the Mind: Embodiment, Action, and Cognitive Extension* Oxford University Press
Clark, A. (1998). *Being There*. MIT Press/Bradford Books, Cambridge, MA
Cohen, Cavanagh, Chun, & Nakayama (2012). The attentional requirements of consciousness. *Trends in Cognitive Sciences* 16: 411–417
Cohen, M.A. and Dennett, D.C. (2011). Consciousness cannot be separated from function. *Trends Cogn. Sci.* 15: 358–364
Craver, C.F. (2007). *Explaining the Brain: Mechanisms and the Mosaic Unity of Neuroscience*. Oxford University Press, New York
Crick, F. (1996). Visual perception: Rivalry and consciousness, *Nature*, 379: 485–6
Christofidou, A.First person: The demand for identification free self-reference. *Journal of Philosophy*, vol. 92 (4): 223–234
Dainton, B. (2000). *Stream of Consciousness: Unity and Continuity in Conscious Experience*. Routledge, London
Danto, A. (1965), *Analytical Philosophy of History* Cambridge University Press, Cambridge
Damasio, A. (2010). *Self Comes to Mind*. Pantheon Books, New York
Damasio, A.R. (2003). *Looking For Spinoza: Joy, Sorrow, and the Feeling Brain* Harcourt Brace, New York
Damasio, A. (1999). *The Feeling of What Happens: Body and Emotion in the Making of Consciousness*. Harcourt Brace & Co., New York
Davies, P. (2009). *Subjects of the World: Darwin's Rhetoric and the Study of Agency in Nature*. University of Chicago Press, Chicago
Dehaene, S., Changeux, J.P., Naccache, L., Sackur, J., & Sergent, C. (2006). Conscious, preconscious, and subliminal processing: a testable taxonomy. *Trends in Cognitive Sciences* 10 (5), 204–211
Deleuze, G. (1994 [1968]). *Difference and Repetition*. The Athlone Press, London
Dennett, D.C. (1992) The self as the center of narrative gravity. In *Self and Consciousness: Multiple Perspectives*, (ed.) Kessel, F.S. Earlbaum, Hillsdale, NJ

Dennett, D.C. (1975). Why the law of effect will not go away. *Journal of the Theory of Social Behavior*, 169–176

Depraz, N. (2008). The rainbow of emotions: at the crossroads of neurobiology and phenomenology. *Cont Philos Rev* 41:237–259

Derrida, J. (1973). *Speech and Phenomena and Other Essays*. Allison, trans., Northwestern University Press, Evanston

Descartes R. (2008) *Meditations on First Philosophy*. Oxford University Press, London

Dobzhansky, T. (1973). Nothing in biology makes sense except in the light of evolution. *American Biology Teacher*, 35: 125–129

Dretske F. (1995) *Naturalizing the mind*. MIT Press, Cambridge, MA

Dreyfus, H.L. (2012). Why Heideggerian AI failed and how fixing it would require making it more Heideggerian. In Kiverstein & Wheeler (eds) *Heidegger and Cognitive Science*. Palgrave Macmillan, Hampshire UK

Dreyfus, H.L. (1992). *What Computers Still Can't Do*. MIT Press, Cambridge, MA

Edelman, G. & Tononi, G. (2000). *A Universe of Consciousness: How Matter Becomes Imagination*. Basic Books, New York

Ereshfsky, M. (2012). Homology Thinking. *Biology and Philosophy*. v.24, no.1, DOI 10.1007/s10539-012-9313-7

Ereshfsky, M. (2007). Psychological categories as homologies: Lessons from ethology. *Biology and Philosophy*. DOI 10.1007/s10539-007-9091-9

Evans, G. (1982). *The Varieties of Reference*. Clarendon Press, Oxford

Farah (2010). Animal neuroethics and the problem of other minds. In Farah (ed.) *Neuroethics*. MIT Press, Cambridge MA

Flanagan, O. (2003). Ethical expressions: why moralists scowl, frown and smile. In Hodge, J. & Radick, G. (eds) *The Cambridge Companion to Darwin*. Cambridge University Press

Flanagan, O., (1992). *Consciousness Reconsidered*. MIT Press, Cambridge

Fodor, J. (1978). Propositional attitudes. *The Monist* 61 (October): 501–23

Freeman, W. (2008). Perception of time and causation through the kinesthesia of intentional action. *Integrative Psychological & Behavioral Science* 42 (2): 137–143

Freeman, W. (2001). *How Brains Make Up Their Minds*. Columbia University Press, New York

Freud, S. (1953–1974). *The Standard Edition of the Complete Psychological Works of Sigmund Freud* 24 volumes, J. Strachey & A. Freud, (eds), Hogarth Press & the Institute of Psycho-analysis, London

Freud, S. (1984 [1925]). The unconscious. In A. Richards, (ed.), *The Penguin Freud Library: Vol 2. On metapsychology*. 159–222, Penguin Books, London

Freud, S. (1953–1974 [1915]). The Unconscious. In *The Standard Edition* 14: 159–216

Gallagher, S. (2008). How to undress the affective mind. *Journal of Consciousness Studies* 15 (2): 89–119

Gallagher, S. (2005). *How the Body Shapes the Mind*. Oxford/Clarendon Press, New York

Gallagher, S. and Varela, F. (2003). Redrawing the map and resetting the time: Phenomenology and the cognitive sciences. In Crowell, Embree, Julian (eds), *The Reach of Reflection: Issues for Phenomenology's Second Century*. Center for Advanced Research in Phenomenology, Inc.

Gallagher & Zahavi (2012). *The Phenomenological Mind* 2nd edition. Routledge Press, London

Geach P. (1973). Ontological relativity and relative identity. In Munitz M.K. (ed.) *Logic and Ontology*. New York University Press, New York
Georgalis, N. (2005). *The Primacy of the Subjective*. MIT Press, Cambridge, MA
Gerhart & Kirschner (2007). The theory of facilitated variation. *PNAS*. 104 (suppl. 1): 8582–8589
Giacino, Fins, Laureys & Schiff (2014). Disorders of consciousness after acquired brain injury: the state of the science. *Nature Reviews Neurology* 10, 99–114
Gibson, J. J. (1979). *The Ecological Approach to Visual Perception*. Houghton Mifflin, Boston
Giere R.N. (1999). *Science without laws*. University of Chicago Press, Chicago
Gillett, G. (1987). The generality constraint and conscious thought. *Analysis* 47: 20–24
Gloor, P. (1997). *The Temporal Lobe and Limbic System*. Oxford University Press, London
Gloor, P. (1994). Is Berger's dream coming true? *Electroencephalography and Clinical Neurophysiology*, 90: 253–266
Glymour, C. (1994). On the methods of cognitive neuropsychology. *British Journal for the Philosophy of Science*. 45: 815–45
Godfrey-Smith, P. (2003). *Theory and Reality: An Introduction to the Philosophy of Science*. University of Chicago Press, Chicago
Graham, G. & Neisser, J. (2000). Probing for relevance: What metacognition tells us about the power of consciousness. *Consciousness and Cognition*, 9 (2): 172–177
Greene, J. (2003). From neural "is" to moral "ought": What are the moral implications of neuroscientific moral psychology? *Nature Reviews Neuroscience* 4: 847–850
Griffiths, P. (2007). Evo-devo meets the mind: Towards a developmental evolutionary psychology. In Sanson & Brandon, (eds)
Grush, R (2006). How to, and how *not* to, bridge computational cognitive neuroscience and Husserlian phenomenology of time consciousness. *Synthese* 153: 417–450
Haybron D. (2010). *The pursuit of unhappiness: the elusive psychology of well-being*. Oxford University Press, New York
Hacking, I. (1998). *Rewriting the Soul: Multiple Personality and the Science of Memory*. Princeton University Press: Princeton, NJ
Hacking, I. (1983). *Representing and Intervening: Introductory Topics in the Philosophy of Natural Science*. Cambridge University Press, New York
Hellie B. (2007). That which makes the sensation of blue a mental fact: Moore on phenomenal relationism. *European Journal of Philosophy* 15 (3): 334–366
Hilgard, E. (1987). *Psychology in America: A historical survey*. FL: Harcourt Brace Jovanovich Publishers, New York & Orlando
Hirst W, Phelps E, et al (2009) Long-term memory for the terrorist attack of September 11: flashbulb memories, event memories, and the factors that influence their retention. *J Exp Psych Gen* 138 (2): 161–176
Horselenberg, R., Merckelback, H., and Josephs, S. (2003), Individual differences and false confessions: A conceptual replication of Kassin and Kiechel, *Psychology, Crime, and Law*, vol. 9:1–8
Howhy, J. (2013). *The Predictive Mind*. Oxford U. Press, New York
Howhy, J. (2012). Neural correlates and causal mechanisms. *Consciousness and Cognition* 21: 691–692

Howhy, J. (2009). The neural correlates of consciousness: new experimental approaches needed? *Consciousness and Cognition.* 18 (2): 428–38

Hutto & Myin (2012). *Radicalizing Enactivism: Basic Minds without Content.* MIT Press, Cambridge, MA

Ikemoto, S. (2010a) Brain reward circuitry beyond the mesolimbic dopamine system: A neurobiological theory, *Neuroscience & Biobehavioral Reviews,* 35 (2): 129–150

Ikemoto (2010b). Reward is exploration. Talk delivered at the Panksepp Festschrift Symposium, Bowling Green University 5/22/10

Ishmael, J.T. (2007). *The Situated Self.* Oxford University Press, New York

Izhikevich, E. M. (2007). *Dynamical Systems in Neuroscience.* MIT Press, Cambridge, MA

Jackendoff, R. (1987). *Consciousness and the Computational Mind.* MIT Press, Cambridge, MA

Jackson, F. (2003). Mind and Illusion. In Anthony O'Hear (ed.), *Minds and Persons.* 421–442, Cambridge University Press

Jackson, F. (1994). Finding the mind in the natural world. Reprinted in Chalmers, (ed.), *Philosophy of Mind: Classical and Contemporary Readings.* New York: Oxford Press

Jackson, F. (1982). Epiphenomenal qualia. *Philosophical Quarterly* 32: 127–136

Jacobs, L. (2012). From chemotaxis to the cognitive map. *Proceedings of the National Academy of Sciences,* 109, (suppl. 1): 10693–10700

Jacobs, L. (2006). From movement to transitivity: The role of hippocampal maps in configural learning. *Reviews in Neuroscience*: 17: 99–109

Jacobs, L. (1996). The economy of winter: Phenotypic plasticity in behavior and brain structure. *Biol. Bull.* 191: 92–100

Jacobs, L & Schenk, F (2003). Unpacking the cognitive map: The parallel map theory of hippocampal function. *Psych. Rev Vol.* 110 (2): 285–315

Kassin, SM. and Kiechel, KL. (1996), The social psychology of false confessions: Compliance, internalization, confabulation, *Psychological Science,* 7: 125–128

Kihlstrom, J.F., Mulvaney, S., Tobias, B.A., & Tobis, I.P. (2000). The emotional unconscious, in E. Eich, ed., *Cognition and Emotion* Oxford University Press, New York

Kitcher, P. (1992). *Freud's Dream: A Complete Interdisciplinary Science of the Mind.* MIT Press, Cambridge, MA

Koch, C. (2004). *The Quest for Consciousness: A Neurobiological Approach.* Englewood, CO: Roberts & Co.

Koriat, A. (2000). The feeling of knowing: Some metatheoretical implications for consciousness and control. *Consciousness & Cognition,* 9: 149–171

Kriegel, U. (2009). *Subjective Consciousness: A self-representational theory.* Oxford University Press, London

Kriegel, U. (2006a) Philosophical theories of consciousness: Contemporary Western perspectives. In *Cambridge Handbook of Consciousness* (Moscovitch, M, Thompson, E., & Zelato, P., Eds). Cambridge University Press, New York

Kriegel, U. (2006b). The Concept of Consciousness in the Cognitive Sciences: Phenomenal Consciousness, Access Consciousness, and Scientific Practice. In P. Thagard (ed.), *Handbook of the Philosophy of Psychology and Cognitive Science.* North-Holland, Amsterdam

Kriegel U. (2005) Naturalizing subjective character. *Philosophy and Phenomenological Research* Vol.LXXI, (1): 23–57

Kripke, S. (1972). Naming and Necessity. In Davidson, Donald and Harman, Gilbert, (eds), *Semantics of Natural Language*. Dordrecht: Reidel: 253–355, 763–769
Kuhl, P.K. (2007). Is speech learning "gated" by the social brain? *Developmental Science*, 10, 110–120
Lakoff, G. & Johnson, M. (2000). *Philosophy in the Flesh* Basic Books, New York
Lambie & Marcel (2002). Consciousness and the varieties of emotion experience: a theoretical framework, *Psycological Review*, vol. 109(2): 219–259
Lamme, V. (2007). Sue Ned Block! Making a better case for p-consciousness. *Behavioral and Brain Sciences*. 30: 481–548
Lamme, V. (2006). Towards a true neural stance on consciousness. *Trends in Cognitive Sciences* 10 (11): 511–512
LeDoux, J. (2002). *Synaptic Self*. Viking Press, New York
LeDoux, J. (1998), Fear and the brain: where have we been, and where are we going? *Biological Psychiatry*, 44: 1229–1238
Leopold, D.A. & Logothetis, N.K. (1996). Activity Changes in early visual cortex reflect monkeys' percepts during binocular rivalry. *Nature*, 379: 549–553
Levine J. (2007) Phenomenal concepts and the materialist constraint. In: Alter T, Walter S *Phenomenal concepts and phenomenal knowledge: New essays on consciousness and physicalism*. Oxford University Press, New York
Levine, J. (2001). *Purple Haze: The Puzzle of Consciousness*.: Oxford University Press, New York
Lewens (2005), What is Darwinian naturalism? *Biology and Philosophy* 20: 901–912
Lewis, D. (1973). *Counterfactuals*. Harvard University Press, Cambridge, MA
Lewontin, R. C. (1970). "The Units of Selection", *Annual Review of Ecology and Systematics* 1, 1–18
Lloyd, D. (2012). Neural correlates of temporality: default mode variability and temporal awareness. *Consciousness and Cognition*, 21(2): 695–703
Lloyd, D. (2002). Functional MRI and the Study of Human Consciousness. *Journal of Cognitive Neuroscience* 14 (6): 818–831
Loar B. (1997) Phenomenal states. In: Block N, Flanagan O, Guzeldere G (eds) *The Nature of Consciousness*. MIT Press, Cambridge, MA
Locke J. (1995[1693]) *An Essay Concerning Human Understanding*. Prometheus Books, Amherst, NY
Love, (2007). Functional homology and homology of function: Biological concepts and philosophical consequences. *Biology and Philosophy*, 22: 691–708
Lutz, A., Dunne J.D. & Davidson, R.J. (2007). Meditation and the neuroscience of consciousness. In P.D. Zelazo, Morris Moscovitch & Evan Thompson (eds), *Cambridge Handbook of Consciousness*. Cambridge
Lutz, A., & Thompson, E. (2003). Neurophenomenology Integrating Subjective Experience and Brain Dynamics in the Neuroscience of Consciousness. *Journal of Consciousness Studies*, 10, (9–10): 31–52
Lycan W. (2001) The case for phenomenal externalism. In: J. Tomberlin (ed.) *Metaphysics*. Blackwell, Oxford
Lycan W. (1997) Consciousness as internal monitoring. In: Block N, Flanagan O, Guzeldere G (eds) *The nature of consciousness*. MIT Press, Cambridge, MA
Lycan, W. (1996). *Consciousness and Experience*. MIT Press, Cambridge, MA
Mackie, J.L. (1974, [1980]). *The Cement of the Universe: A Study of Causation*. Clarendon Press, Oxford & New York

MacLean PD (1952) Some psychiatric implications of physiological studies on frontotemporal portion of limbic system (visceral brain). *Electroencephalography and Clinical Neurophysiology,* 4: 407–418

MacLean P.D. (1973). *A Triune Concept of the Brain and behavior.* University of Toronto Press, Toronto

Mandik, P. (2009). The neurophilosophy of subjectivity. In Bickle, Ed. *The Oxford Handbook of Philosophy and Neuroscience.* Oxford University Press, New York

Marabou, C. (2012). *The New Wounded: From neurosis to brain damage.* Miller, trans. Fordham University Press, New York

Matthews G.B. (1977) Consciousness and Life. *Philosophy* 52(199): 13–26

Maturana, H.R. & Varela, F.J. (1980). *Autopoiesis and Cognition: The Realization of the Living.*Boston: D. Reidel Publishing Company, Boston

Mayr E (1959). Typological versus Population Thinking. In *Evolution and Anthropology: a Centennial Appraisal.* 409–412, The Anthropological Society of Washington, Washington D.C.

McDowell J. (1994) *Mind and World.* Harvard University Press, Cambridge, MA

Mendelson, J (1972). Ecological modulation of brain stimulation effects. *Int J Psychobiol* 1: 285–304

Merleau-Ponty, M. (1962). *Phenomenology of Perception.* Smth, trans. Routledge Press, London

Metzinger, T. (2004). Appearance is not knowledge: The incoherent straw man, content-content confusions and mindless conscious subjects. *Journal of Consciousness Studies,*11(1), 67–72

Metzinger, T. (2003a). The "subjectivity" of subjective experience: A representationalist analysis of the first-person perspective *Networks* 3–4: 33–64

Metzinger, T. (2003b). *Being No One.* MIT Press, Cambridge, MA

Metzinger, T. (2000). *Neural Correlates of Consciousness.* MIT Press, Cambridge, MA

Mitra, P. & Bokil, H. (2008) *Observed Brain Dynamics.* Oxford University Press, New York

Moore, G.E. (1997 [1903]) *Principia Ethica 2nd Edition.* Baldwin, (Ed.) Cambridge University Press

Moore G.E. (1903) The refutation of idealism. *Mind* 12: 433–53

Moran R. (2001) *Authority and estrangement: an essay on self-knowledge.* Princeton University Press, Princeton, NJ

Moran R. (1999). The authority of self-consciousness. *Philosophical Topics* 26(1&2): 179–200

Murthy, VN (2011).Olfactory maps in the brain. *Annu Rev Neurosci* 34: 233–258

Nagarjuna, & Garfield, J. L. (1995). *The Fundamental Wisdom of the Middle Way: Nagarjuna's Mulamadhyamakakarika* (J. L. Garfield, Trans.). New York, Oxford: Oxford University Press (Translation and commentary by J. L. Garfield)

Nagel, T. (1974). What is it like to be a bat? *Philosophical Review,* 83: 435–450

Nahmias, E. (2002) Verbal reports on the contents of consciousness: Reconsidering introspectionist methodology. *Psyche* 8(21)

Neander, K.L. (2002). Types of Traits: The Importance of Functional Homologues, in *Functions: New Readings in the Philosophy of Psychology and Biology,* edited by Andre Ariew, Robert Cummins and Mark Perlman. Oxford University Press

Neisser, J. (2014) What subjectivity is not. *Topoi: An International Review of Philosophy* DOI: 10.1007/s11245-014-9256-5

Neisser, J. (2012a), Neural correlates of consciousness reconsidered. *Consciousness and Cognition* 21 (2012) 681–690

Neisser, J. (2012b). Neural mechanisms and functional realization. *Consciousness and Cognition* 21, 693–694

Neisser, J. (2008). Subjectivity and the limits of narrative. *Journal of Consciousness Studies* 15, no.2, 51–66

Neisser, J. (2006a). Unconscious subjectivity. *Psyche: An Interdisciplinary Journal of Research on Consciousness*, 12 (3): 1–14

Neisser, J. (2006b). Making the case for unconscious feeling. *Southwest Philosophy Review* 22 (1): 129–138

Neisser, J. (2005). Psychoneural reduction and the future of psychology. *Phenomenology & The Cognitive Sciences* 4 (3): 259–269

Neisser, J. (2003). The swaying form: Metaphor, imagination, embodiment. *Phenomenology & The Cognitive Sciences*, 2, (1): 27–53

Neisser, J. (1999). On the use and abuse of Dasein in cognitive science. *Monist* 82 (2): 347–361

Neisser (1997), *The Embodied Approach to Cognition: A Defense* Ph.D. dissertation, Duke University

Neisser, U. & Becklen, R. (1975). Selective looking: Attending to visually specified events. *Cognitive Psychology* 7 (4): 480–494

Noe, A. (2009). *Out of Our Heads: Why You Are Not Your Brain, and Other Lessons from the Biology of Consciousness.* Hill and Wang Publishers, New York

Noe, A. (2006) Precis of *Action in Perception. Psyche: An interdisciplinary Journal of Research on Consciousness* 12(1): 1–35

Noe, A. (2004). *Action in Perception.* MIT Press, Cambridge, MA

Noe, A. & Thompson, E. (2004a). Are there neural correlates of consciousness? *Journal of Consciousness Studies*,11(1): 3–28

Noe, A. & Thompson, E. (2004b). Sorting out the neural basis of consciousness: Authors' reply to commentators. *Journal of Consciousness Studies*,11(1): 87–98

Noggle, R. (1998), Review of The Constitution of Selves, *Ethics*, vol. 108, no. 4: 802–805

Nussbaum, M (2001). *Upheavals of Thought: The Intelligence of Emotions* (New York: Cambridge University Press)

O'Doherty, J. (2012). Beyond simple reinforcement learning: the computational neurobiology of reward-learning and valuation. *European Journal of Neuroscience*, 35: 987–990

Owen, R. (1848). *On the Archetype and Homologies of the Vertebrate Skeleton*, John van Voorst, London

Owen R and Cooper, W.W. (1843) *Lectures on the Comparative Anatomy and Physiology of the Invertebrate Animals, Delivered at the Royal College of Surgeons, in 1843.* Longman, Brown, Green, and Longmans, London

Pabst DA (2000). To bend a dolphin: convergence of force transmission designs in cetaceans and scombrid fishes. *Am Zool* 40: 146–155

Panksepp, J. (2007) Affective consciousness, in Velmans, M. & Schneider, S. (eds) *The Blackwell Companion to Consciousness*, 114–129, Blackwell, Malden, MA

Panksepp, J. (2005). Toward a science of ultimate concern. *Consciousness and Cognition* 14: 22–29

Panksepp, J. (2005b). Affective consciousness: Core emotional feelings in animals and humans. *Consciousness and Cognition* 14(1): 30–80

Panksepp, J. (1998). *Affective Neuroscience: The Foundations of Human and Animal Emotions.* Oxford University Press, New York

Panksepp, J., and Biven, L. (2012). *The Archaeology of Mind: Neuroevolutionary Origins of Human Emotion.* W. W. Norton & Company, New York

Panksepp, J., Asma, S. Curran, G., Gabriel, R., & Greif, T. (2012). The Philosophical Implications of Affective Neuroscience. *Journal of Consciousness Studies,* 19 (3–4): 6–48

Papineau D. (1993) *Philosophical Naturalism.* Blackwell, Oxford, UK

Papineau D. (2002) *Thinking about Consciousness.* Oxford University Press, New York

Parfit D. (1999) Experiences, subjects, and conceptual schemes. *Philosophical Topics* 26(1&2): 217–270

Parfit, D. (1984), *Reasons and Persons.* Clarendon Press, Oxford

Pearl, J. (2000). *Causality: Models, Reasoning and Inference.*: Cambridge University Press, New York

Pennartz, Ito, Verschure, Battaglia, & Robbins (2011). The hippocampal-striatal axis in learning, prediction and goal-directed behavior. *Trends in Neurosciences* v.34 (10): 548–559

Perry, J. (2002). *Identity, Personal Identity, and the Self* : Hackett Publishing, Indianapolis and Cambridge

Perry, J. (1993). *The Problem of the Essential Indexical and Other Essays.* Oxford University Press

Perry, John, (1986). Thought without Representation. *Proceedings of the Aristotelian Society* Supplementary Volume 60: 263–283

Petitot, Varela, Pachoud, & Roy (1999) *Naturalizing Phenomenology: Issues in Contemporary Phenomenology and Cognitive Science*: Stanford University Press, Palo Alto, CA

Place U.T. (1956) Is consciousness a brain process? *British Journal of Psychology* 47: 44–50

Prinz, J. (2012). *The Conscious Brain: How Attention Engenders Experience.* Oxford University Press, New York

Prinz, J. (2005) Are emotions feelings? *Journal of Consciousness Studies,* 12 (8–10): 9–25

Pryor J. (1999) Immunity to error through misidentification. *Philosophical Topics* 26 (1&2): 271–304

Putnam, H. (1973). Psychological predicates. In Capitan & Merril (eds) *Art, Mind, and Religion.* University of Pittsburgh Press, Pittsburgh, PA

Ramachandran, V.S. and Oberman, L.M. (2007). Broken mirrors: A theory of autism. *Scientific American,* 17 (20–29)

Reed, E. (1997). *From Soul to Mind: The Emergence of Psychology, from Erasmus Darwin to William James.* Yale University Press: New Haven, Conn.

Ricoeur, P. (1984), *Time and Narrative I,* trans. K. McLaughlin and D. Pellauer. University of Chicago Press, Chicago

Ricoeur, P. (1985), *Time and Narrative II*, trans. K. McLaughlin and D. Pellauer. University of Chicago Press, Chicago
Ricoeur, P. (1988), *Time and Narrative III*, trans. K. Blamey and D. Pellauer. University of Chicago Press, Chicago
Recanati, F. (2006). Crazy Minimalism. *Mind and Language*, 21
Reep, Finlay, & Darlington (2007). The Limbic System in Mammalian Brain Evolution. *Brain Behav Evol*, 70: 57–70
Rees, G., Krieman, G. and Koch, C. (2002). Neural correlates of consciousness in humans, *Nature Reviews Neuroscience*, 3:261–70
Rodriguez, E., George, N., Lachaux, J.P., Martinerie, J., Renault, B. & Varela, F.J. (1999). Perception's long shadow: long distance synchronization of human brain activity. *Nature* 397: 430–433
Rorty, R. (1991), Freud and moral reflection. In *Essays on Heidegger and Others: Philosophical Papers vol. 2* (New York: Cambridge University Press)
Rorty R. (1970) Incorrigibility as the mark of the mental. *The Journal of Philosophy* 67 (12): 399–424
Rorty R. (1974) More on incorrigibility. *Canadian Journal of Philosophy* 4 (1): 195–197
Rosenthal, D.M (2002b). Consciousness and Higher-Order Thoughts. In L. Nadel (ed.), *Macmillan Encyclopedia of Cognitive Science*. Macmillan Publishers, New York
Rosenthal, D.M. (2000). Consciousness and Metacognition, In D. Sperber (ed.), *Metarepresentation*. Oxford UP, Oxford
Rovane, C. (1993), Self-reference: The radicalization of Locke, *Journal of Philosophy*, XC 2: 73–97
Russell B. (1911) Knowledge by acquaintance and knowledge by description. *Proceedings of the Aristotelian Society, New Series* XI (1910–11) 108–128
Russell B. (1914) *Our knowledge of the external world as a field for scientific method in philosophy*. Open Court publishing, Chicago
Sansom & Brandon, eds. (2007). *Integrating Evolution and Development*. Cambridge, MA : MIT Press (A Bradford Book)
Sarnet, HB & Netsky, MG. (1981). *Evolution of the Nervous System* (2nd ed). Oxford University Press, Oxford, UK
Sartre, JP. (1956), *Being and Nothingness*, trans. H. Barnes. Washington Square Press, New York
Searle, J. (2004). Peer Commentary on *Are There Neural Correlates of Consciousness*. *Journal of Consciousness Studies* 11 (1): 80–82
Schechtman, M. (1992), *The Constitution of Selves*. Cornell University Press, Ithaca, NY
Schleiermacher, F (2011 [1830]), *The Christian Faith*. Apocryphile Press, Berkeley, CA
Sellars W. (1997 [1956]) *Empiricism and the philosophy of mind*. Harvard Press, Cambridge, MA
Sellars, W. (1963) Philosophy and the scientific image of man. In *Frontiers of Science and Philosophy*, Colodny, R. (ed.). University of Pittsburgh Press, Pittsburgh 1962: 35–78. Reprinted in *Science, Perception and Reality*
Sharot T, Delgado M, Phelps E (2004) How emotion enhances the feeling of remembering. *Nature Neuroscience* 7(12) 1376–1380

Schear (2013). *Mind, Reason, and Being-in-the-World: The McDowell-Dreyfus Debate*. Routledge, London

Schiff & Fins (2003). Hope for "comatose" patients. *Cerebrum* 5: 7–24

Sheinberg, D.L. & Logothetis, N.K. (1997). The role of temporal cortical areas in perceptual organization. *Proceedings of the National Academy of Sciences*, 94: 3408–3413

Shin, K. et al (2009). The effect of spatial attention on invisible stimuli. *Atten. Percept. Psychophy.* 71: 1507–1513

Shoemaker, S. (1997). The first-person perspective. In Block et al (eds) *The Nature of Consciousness*. MIT Press, Cambridge, MA

Shoemaker, S. (1984). *Identity, Cause, and Mind*. Cambridge University Press, Cambridge)

Shoemaker, S. (1981). Some varieties of functionalism. *Philosophical Topics* 12: 93–119

Shoemaker, S. (1968). Self-reference and self-awareness. *Journal of Philosophy* 65 (19): 555–567

Shutter, D.J.L.T., Van Honk, A. & Panksepp, J. (2004). Introducing transcranial magnetic stimulation (TMS) and its property of causal inference in investigating brain-function relationships. *Synthese* 141: 155–73

Singer, W. (1995). Putative functions of temporal correlations in neocortical processing, in *Large-Scale Neuronal Theories of the Brain*, (ed.) Christof Koch and Joel L. Davis, The MIT Press/A Bradford Book, Cambridge, MA

Smart J.J.C. (1959). Sensations and brain processes. *Philosophical Review* 68: 141–156

Smith, J.D. (2012). The Highs and Lows of Theoretical Interpretation in Animal-Metacognition Research. *Philosophical Transactions of the Royal Society, Biological Sciences* vol. 367, (1594): 1297–1309

Sober, E. (1998). The genetic fallacy. *Routledge Encyclopedia of Philosophy*. Routledge: London

Solms, M. (2003). *The Brain and the Inner World: An Introduction the Neuroscience of Subjective Experience*. Other Press: New York

Solms, M. & Panksepp, J. (2012). The "Id" Knows More than the "Ego" Admits: Neuropsychoanalytic and Primal Consciousness Perspectives on the Interface Between Affective and Cognitive Neuroscience. *Brain Sci. 2* (2): 147–175

Spivey, M. (2007). *The Continuity of Mind*. Oxford University Press, New York

Steinbock (1995). *Home and Beyond: Generative Phenomenology after Husserl*. Northwestern University Press, Evanston

Strausfeld & Hirth (2013). Deep Homology of Arthropod Central Complex and Vertebrate Basal Ganglia. *Science* 340: 157–161

Strawson, G. (2009). *Selves*. Oxford University Press, New York

Strawson G. (2006). *Consciousness and Its Place in Nature: Does Physicalism Entail Panpsychism?* Imprint Academic, Charlottesville, VA

Strawson, G. (2004), Against narrativity, *Ratio (new series)*, XVII 4: 428–452

Sturgeon S. (1994) The epistemic view of subjectivity. *Journal of Philosophy* 91: 221–235

Suppes, P. (1967). What is a scientific theory? In Sidney Morgenbesser (ed.) *Philosophy of Science Today*. 55–67. Basic Book Inc., New York

Syal & Finlay (2011). Thinking outside the cortex: social motivation in the evolution and development of language. *Developmental Science* 14 (2): 417–430

Teichert, D. (2004), Narrative, identity, and the self, *Journal of Consciousness Studies*, 11: 175–191
Thompson, E. (2007). *Mind in Life: Biology, Phenomenology, and the Science of the Mind*. Belknap/Harvard Press, Cambridge, MA
Thompson, E., Lutz, A., & Cosmelli, D. (2005). Neurophenomenology: An Introduction for Neurophilosophers. In Andy Brook and Kathleen Akins (eds), *Cognition and the Brain: The Philosophy and Neuroscience Movement*. Cambridge University Press, New York and Cambridge
Tomasello, M (2014*). A Natural History of Human Thinking*. Harvard Press, Cambridge, MA
Tononi, G. (2012). *Phi: A Voyage from the Brain to the Soul*. Pantheon, New York
Tye M. (2009). *Consciousness revisited: materialism without phenomenal concepts*. MIT Press, Cambridge, MA
Tye, M. (2005 [2003]). *Consciousness and Persons: Unity and Identity*. MIT Press, Cambridge, MA
Tye M. (2000). *Consciousness, Color, and Content*. MIT Press, Cambridge, MA
Tye M. (1995). *Ten Problems of Consciousness*. MIT Press, Cambridge, MA
Van den Bussche, E. et al (2010). The relation between consciousness and attention: An empirical study using the priming paradigm. *Consciousness and Cognition* 19: 86–97
Van Gelder, T. (1995). What Might Cognition Be, If Not Computation? *Journal of Philosophy*, 91 (7): 345–81
Van Fraassen, B. (1980). *The Scientific Image*. Oxford University Press, Oxford & New York
Vanderweele & Robins (2009). Minimal sufficient causation and directed acyclic graphs. *Annals of Statistics,* 37 (3): 1437–1465
Varela, F. J. (1996). Neurophenomenology. A methodological remedy for the hard problem. *Journal of Consciousness Studies*, 3 (4): 330–349
Varela, F.J., Thompson, E.T., & Rosch, E. (1992). *The Embodied Mind: Cognitive Science and Human Experience*. The MIT Press
Vimal, R. (2009). Dependent Co-origination and Inherent Existence: Dual-Aspect Framework. *Vision Research Institute: Living Vision and Consciousness Research* [Available: http://sites.google.com/site/rlpvimal/Home/2009-Vimal-Cooriginatíon-LVCR-2-7.pdf], 2 (7): 1–50
Von Eckardt, B. (1993). *What is Cognitive Science?* Bradford Books/M.I.T. Press, Cambridge, MA
Wagner G.P. (2007). The developmental genetics of homology. *Nature Reviews Genetics* 8: 473–479
Wagner, G.P. (1996). Homologues, Natural Kinds and the Evolution of Modularity, *American Zoologist* 36: 36–43
Williams, C. (1998). Review of The Constitution of Selves, *The Philosophical Review*, 107 (4): 641–644
Wilson A.D. and Golonka S. (2013.) Embodied cognition is not what you think it is. *Front. Psychology* 4: 58
Wilson, E.O. (1998). *Consilience: The Unity of Knowledge*. Alfred A. Knopf, New York
Wimsatt, W.C. (1986). Developmental constraints, generative entrenchment, and the innate-acquired distinction, in P. W. Bechtel. (ed.) *Integrating Scientific Disciplines*. 185–208, Martinus-Nijhoff, Dordrecht

Wimsatt, W.C. & Griesemer, J. (2007). Reproducing Entrenchments to Scaffold Culture: The Central Role of Development in Cultural Evolution. In *Integrating Evolution and Development: From Theory to Practice*, Sansom & Brandon (eds), 228–323. MIT Press, Cambridge, MA

Wittgenstein, L. (1958). *Preliminary Studies for the "Philosophical Investigations," Generally Known as the Blue and Brown Books* Basil Blackwell, Oxford

Zahavi, D. (2014). *Self and Other: Exploring Subjectivity, Empathy, and Shame*. OUP

Zahavi, D. (2005). *Subjectivity and Selfhood*. MIT Press, Cambridge, MA

Zahavi, D. (1999). *Self-awareness and Alterity: A Phenomenological Investigation*. Northwestern University Press, Evanston, IL

Index

a priori physicalism, 111–12, 114, 138, 179nn. 3–4
access consciousness (a-consciousness), 41–6, 50
affective ignorance, 43
affective neuroscience
  active-organism view and, 102
  affective consciousness and, 101, 106, 176n. 1, 179n. 25
  versus behavioral neuroscience, 101–2
  as dual aspect monism, 104–5, 178n. 22
  evolutionary developmental biology (evo-devo) and, 4, 107–8
  limbic system and, 58
  modular embodiment and, 111
  place-preference and, 103
  primary affective modes and, 101–8
  unconscious subjectivity and, 57–8, 104–8
affective salience, 110, 135–6, 138
AIR model of consciousness ((A)ttended (I)ntermediate level (R)epresentations), 52
Alcaro, Antonio, 102
Ammon's horn, 95
amygdala, 92–4, 159
animal navigation, *see* navigation
Anscombe, G.E.M., 18, 165
anterior cingulate cortex, 106
argument by analogy (Mill), 105, 178n. 24
argument from neural homology (Panksepp), 105
Aristotle, 27, 34
Atkins, Kim, 26–9, 34–5, 168n. 32, 168n. 37, 168n. 41, 169n. 63, 170n. 64
attention, 17, 50–3
autopoiesis (self-organization), 7–8, 90, 124, 128, 162, 186n. 7

awareness, *see also* consciousness; inner awareness
  attention and, 51
  first-person perspective and, 71
  focal awareness and, 44
  peripheral awareness and, 44–5
  qualia and, 65–6, 68–73
  subjectivity and, 19, 40–1, 65–6, 131
  transitive nature of, 182n. 33

backmasked stimuli, 49–50
baseline subjectivity, *see also* subjectivity
  across diverse ecological and behavioral contexts, 4, 25, 88–9, 117, 123
  affective salience and, 110, 135–6, 138
  dynamic limbic systems and, 113
  first-person unconscious and, 56
  identification free, affectively valenced spatiotemporal orientation and, 25, 113
  prereflective self-consciousness (PRSC) and, 131, 182n. 33
Bearing Map, 95–6, 98, 108, 135, 137
behavioral neuroscience, 100–3, 105, 117, 120
Berkeley, Bishop George, 67–9, 78, 163
Berra, Yogi, 15, 166n. 6
Berridge, Kent, 137–8
Bickle, John, 157
binding problem, 154
binocular rivalry paradigm, 5, 142–3, 146–52, 156
blindsight, 45–6, *see also* superblindsight
Block, Ned, 41–51, 171n. 9, 174n. 18
body-body problem (Thompson), 7
Brandom, Robert, 86
Brandon, R.N., 117
Bruner, Jerome, 86

Campbell, John, 16, 22–3, 36–7, 166n. 12, 169n. 53
CARE (primary affective mode), 103
Carruthers, Peter, 19–20
central state identity theory, 133, 174n. 21, 180n. 5, 183n. 42
cerebral unconscious, 57
Chalmers, David, 63, 112, 140–1, 143–8, 150–1, 154, 167n. 20, 173n. 11, 183n. 1, 183–4n. 2, 184nn. 4–5
change blindness, 51
chemical-visceral system, 92, see also limbic system
Christofidou, Andrea, 32–3
Churchland, Patricia Smith, 1, 114–15, 157
Churchland, Paul, 157
cingulate gyrus, 92
Clark, Andy, 123–4, 127, 181n. 19
cognitive maps
　Freeman on, 98–100
　as metaphors, 99–100
　navigation and, 4, 90, 95, 99, 107
　neighborhood avoidance example and, 107
　olfactory spatial hypothesis (OS) and, 95
　salience landscape and, 88
cognitive model of unconscious thought
　emotion and, 55–7, 59
　neuroscience and, 55, 102
　representation concept in, 55–6
　unconscious subjectivity and, 41, 54–5, 57–8, 60
Cohen, Michael A., 50–2
concept location (CL) project in philosophy, 160–2
consciousness, see also awareness; neural correlates of consciousness (NCC) research
　access consciousness (a-consciousness) and, 41–6, 50
　access-only consciousness (a-only consciousness) and, 42, 45–6
　affective ignorance and, 43
　AIR model of, 52
　attention and, 50–3
　central state identity theory and, 133
　change blindness and, 51
　content consciousness and, 141–2, 144–5, 147–8
　contrastive method and, 47
　cortico-thalamic connectivity and, 91, 122
　creature consciousness and, 141, 144–6, 167n. 20
　emotional consciousness and, 52
　explicit versus implicit representation and, 45
　first-person perspective and, 61, 145
　focal awareness versus peripheral awareness and, 44–5
　for-me mental states and, 38–9, 45–7
　"hard problem" of, 38, 55, 61, 64, 133
　holism of conscious content approach to, 20–1, 128, 143–4, 148–9, 153–4, 156
　iconic memory and, 50
　inattentional blindness and, 50–1
　inner awareness and, 45, 47
　local reentrant processing and, 48, 50, 53
　Locke's definition of, 43
　materialist approach to, 63, 65
　metacognition and, 7, 41–3, 45–6
　minimal conscious state (mcs) and, 53
　"mongrel" nature of, 42, 45–7
　narrative subject model and, 28
　neo-Lockean conception of, 44–7, 50
　"now" phase of, 130–1, 134
　phenomenal consciousness (p-consciousness) and, 41–2, 44, 47
　phenomenal-only consciousness (p-only consciousness) and, 42–51, 171nn. 9, 12, 174n. 18
　problem of unreportable experience and, 47–9
　qualia and, 63, 65–6, 68, 74, 155, 162
　representation and, 48, 55

consciousness, *see also* awareness; neural correlates of consciousness (NCC) research – *Continued*
   self-representationalism and, 39, 45–7, 173n. 13
   subjectivity and, 2–3, 6–7, 39–42, 44, 46–7, 52–3, 55, 60, 72, 88, 114, 145, 165
   "superblindsight" and, 45–6
   transcendental phenomenology and, 124, 162
   transparency thesis and, 73–6, 79, 174n. 19, 175–6n. 28
   unconscious subjectivity contrasted with, 38, 41–2
   visual consciousness and, 52, 145, 149, 184n. 3
content consciousness, 141–2, 144–5, 147–8
continuity framework, 127
contrastive method, 47
control theory, 126–7
coping
   as affective orienteering, 87–9
   mapping with affective valence and, 93, 121, 177n. 8
   as motivated learning, 121–2
   navigation and, 90, 93
   subjectivity and, 4, 7, 19, 84, 87–9, 93, 110, 121–2
cortico-thalamic connectivity, 91, 122
Craver, Carl, 151–4, 185n. 8
creature consciousness, 141, 144–6, 167n. 20
Crick, Francis, 140, 149, 183n. 1

Dainton, Barry, 143
Damasio, Antonio, 54
Danto, Arthur C., 27
Darlington, Richard, 94
Darwinism, 84, 121, 158–62, 164
Davies, Paul, 160–2, 186nn. 5, 7
Dennett, Daniel, 50, 169n. 45
dentate gyrus, 95
depth psychology, 56, 59
Descartes, René, 66, 86, 173n. 8
developmental conservation, 122
developmental covariation, 89, 105–6, 108, 118

developmental modules, 120–1, 128, 137, 180–1n. 14
Dobzhansky, Theodosius, 119
dorsal striatum, 103, 108
Dreyfus, Hubert, 123
dual-task paradigm, 49, 51
dynamic systems theory (DST)
   continuity framework and, 127
   embodiment and, 124–7
   functional decomposition and, 121, 126, 134
   model/mechanism distinction and, 125
   mutual feedback and, 127
   neurophenomenology® and, 133–5
   predictive coding and, 126–7, 135–7
   reverse engineering and, 125
   as a topic-neutral description, 124–5, 133

eliminativism, 40, 80, 163–4
embodiment
   dynamic systems theory (DST) and, 124–7
   embodied functionalism and, 123–4, 127
   embodied phenomenology, 99, 123–4, *see also* enactivism
   leib-korper distinction and, 124, 129
   minimal cognitivism and, 127
   modular embodiment and, 25, 111, 123–5, 128–9
   subjectivity and, 1, 4, 110–11, 165
emotion
   cognitive model of the unconscious and, 55–7, 59
   neostoic approach to, 59–60
   neuroscience of, 54, 101
   subjectivity and, 42, 57–60, 71
   unconscious subjectivity and, 42, 56–60, 71
emotional brain, 92, *see also* limbic system
emplotment, 27–8, 31, 35–6
enactivism
   anti-cognitivist theme in, 99
   autopoiesis and, 124, 162
   concept location in philosophy
   critique of, 161

enactivism – *Continued*
  holism of conscious content approach and, 128, 148–9, 153–4
  leib-korper distinction in, 124
  neural correlates of consciousness (NCC) research and, 148–9, 152–3
  neurophenomenology and, 124
  neurophenomenology® and, 134
entorhinal cortex, 91–3, 96–7, 137
epistemic/pragmatic relations, 35
Ereshefsky, Marc, 99, 111, 116, 180n. 8
"esse is *percipi"* (Berkeley), 67, 69, 78
Euclidean value code, 100
eudaimonia, 59–60
Evans, Garreth, 13, 17–20, 22–3, 166nn. 8, 11–12
evolutionary developmental biology (evo-devo)
  affective neuroscience and, 4, 107–8
  cognitive evo-devo and, 89, 122, 138
  developmental covariation and, 89–90, 95, 114, 118
  holism of conscious content approach and, 129
  homology thinking and, 5, 89, 110, 116–22
  mechanistic approach to neuroscience and, 129
  modular embodiment and, 111, 123–5, 128–9
  navigation and, 90, 93
  neural trait polarity and, 89
  olfactory navigation and, 97
  plasticity and, 89
  reverse engineering and, 115–16
  subjectivity and, 8, 93, 110, 118–21
  triune brain hypothesis and, 92
evolutionary psychology, 5, 89, 115–18, 136
explanatory gap, 41, 55, 62, 109, 119, 124, 155–6

false confessions, 30
Farah, Martha, 105–6, 178n. 24
feature binding, 49–50, 52
Fechner, Gustav, 142–3
figure ground segregation, 49–50, 52
fine odor discrimination, 93–4

Finlay, Barbara, 89, 94, 121–2
first-person perspective
  coffee cup example and, 16, 20
  consciousness and, 2
  descriptive reference and, 31
  holistic nature of, 20–1
  identification free self-reference and, 12, 113
  motivated learning and, 121–2
  narrative self and, 169n. 58
  navigation and, 84, 98, 100, 107, 135
  neurobiological basis for, 1, 108–9, 118
  neurophenomenology of, 111
  situated subjectivity and, 20–1
  subjectivity and, 2, 4, 11–12, 20–1, 29, 47, 52, 61, 67, 70, 88, 91, 108, 110, 114, 146, 166n. 11
first-person unconscious, 56–7
Flanagan, Owen, 84–5
focal awareness, 44
for-me subjectivity
  conscious awareness and, 38–9, 45–7
  for-me/in-me distinction and, 2–3
  identification free self-reference and, 13–18, 61
  I-thoughts and, 17–20, 23
  navigation and, 114
  Space of Reasons and, 19
  subject position and, 6, 23
  unconsciousness and, 48, 54
fornix, 92
Freeman, Walter, 97–100, 108–9, 114, 136, 177n. 16, 178nn. 17–18
Freud, Sigmund, 39, 54, 56, 58, 107, 179n. 26
functional decomposition, 121, 126, 134, 162
functionalism, 4–5, 123–4, 127–8, 180n. 10, 183n. 42

Gallagher, Shaun, 123, 129–30, 134, 181n. 28, 182n. 31
genealogy, 158–60, 162, 164
Generality Constraint, 18–19
genetic fallacy, 158–60, 186n. 5
gist perception, 49
global reentrance, 48, 171n. 16

Index 205

gradient maps, 100
Griffiths, Paul, 117

habit, 85
Haybron, Dan, 43
Hegel, G.W.F., 86
Heidegger, Martin, 176–7n. 3
Hellie, Benj, 65, 74, 174n. 19
higher-order thought (HOT) theory, 7, 170n. 2
hippocampus, 91–7, 99–100, 108, 136–8, 181n. 16
holism of conscious content approach, 20–1, 128–9, 143–4, 148–9, 153–4, 156
homologues
 definitions of, 116–17, 180n. 6
 developmental covariation and, 89, 105–6, 108, 118–19
 developmental modules and, 120–1, 128, 137, 180–1n. 14
 iterative homologues and, 181n. 16
 language acquisition/vocal learning example of, 121–2
 mammalian forelimb example of, 117–18, 120–1
 process homologues and, 119–20, 137
 quasi-independence of, 120
 trait polarity and, 118
homology thinking
 argument from neural homology (Panksepp) and, 105
 character/character state distinction and, 116–17
 definition of, 111
 developmental conservation and, 122
 evolutionary developmental biology (evo-devo) and, 5, 89, 110, 116–22
 holism of conscious content approach and, 128
 language acquisition and, 121–2
 lineages and, 117, 128
 population thinking and, 125
Howhy, Jakob, 126
Huber, Robert, 102
humanism, 85, 161

Hume, David, 1, 154, 177n. 8
Husserl, Edmund, 5–6, 111, 129–36, 182n. 33
Hutto, Daniel, 123, 126, 181n. 17
hypothalamus, 92, 102–4, 108, 114,
 see also lateral hypothalamus

"I" pronoun
 objective use of, 12–13, 18, 29–32
 subjective use of, 12–13, 15, 18, 29, 32, 37
iconic memory, 48–50, 52
idem identity, 168n. 41
identification free self-reference
 anterograde amnesia example and, 33–4
 Christofidou on, 32–3
 first-person "mode of presentation" and, 16
 for-me-ness and, 13–18, 61
 immunity to error through misidentification and, 3, 12–14, 174n. 15
 I-thoughts and, 17–18, 23
 Rovane's alternative to, 32–4
 simple perceptual demonstration and, 14, 17
 subjectivity as, 3, 6–7, 12–20, 110, 113, 131, 135, 138, 163–5
immediacy intuition, 63–4, 69–70, 73, 173n. 14
immunity to error through misidentification
 ballpark video example, 12, 15, 30
 Columbus Stockade Blues example and, 14–15
 definitional truths and, 14–15
 identification free self-reference and, 3, 12–14, 174n. 15
 narrative subject model and, 29–33
 objection to the idea of, 21–3
 Russell's theory of descriptions and, 14
in me/for-me distinction, 39, 54–5
inattentional blindness, 50–1
incentive salience, 137–8
inferior temporal cortex (IT), 143, 149, 156

inner awareness doctrine, 13, 24, 39, 45, 47, 61, 66–73, 76, see also qualia
inner psychophysics, 142–3, 145, 152
inner time consciousness, 129, 132, 134
Integrated Map, 96, 98, 108, 177n. 13
intersubjectivity, 17, 86, 90
intrinsic experientiality, 145–6
inus conditions, 150–1, 153, 156
ipse identity, 28–9, 35, 168n. 41, 169n. 63
I-thoughts
 for-me experience and, 17–20, 23
 generality constraint and, 18–19
 identification free self-reference and, 17–18, 23
 Kantian arguments regarding, 33

Jackendoff, Ray, 52
Jackson's Constraint (Frank Jackson), 111–14, 124, 138, 179nn. 2–4
Jacobs, Lucia, 93–100, 108, 177n. 9, 178n. 18
James, William, 160, 182n. 30
joint attention, 17

Kant, Immanuel, 33, 90, 186n. 7
Kassin, S.M., 30
Knowledge Argument, 6, 163
korper (body taken as object of biological science), 124, 129
Kriegel, Uriah, 7, 24, 39–40, 44–5, 47, 77, 170nn. 2, 8, 171nn. 9–10, 173n. 13, 175n. 27

Lamme, Victor, 48, 50–1, 53
language acquisition, 121–2
lateral hypothalamus (LH), 103–4, 108
LeDoux, Joseph, 54, 58–60
leib (lived body), 124, 129
Leibniz, G.W., 15, 69, 160
Leopold, David A., 142
Levine, Joseph, 6–7, 24, 40, 61, 63–8, 71, 75, 78, 80, 110, 112, 114, 155, 173n. 11, 175n. 27
limbic loop model, *see* space-time loop

limbic system
 affective neuroscience and, 58
 developmental covariation of, 121
 identification free, affectively valenced spatiotemporal orientation and, 113
 ML-DA System (mesolimbic / dopaminergic system) and, 91, 102–4, 107, 122, 137
 navigation and, 93, 103–4, 107–8, 113, 136–7
 olfactory-limbic system and, 91, 93
 subjectivity and, 91, 100, 113
 triune brain hypothesis and, 92
local reentrant processing, 48, 50, 53, 171n. 16
Locke, John, 42–4, 68
locked-in syndrome, 53
Logothetis, Nikos, 142–3, 148, 150, 153
Love, Alan C., 119–20

McDowell, John, 19
Mackie, J.L., 150, 152, 185n. 8
MacLean, Paul, 91–2
mamillary bodies, 92
Mandik, Pete, 157, 163
manifest image, 83–5, 158, 164–5
Marabou, Catherine, 56–7
Maturana, Humberto, 128, 153–4, 181n. 26
Mayr, Ernst, 115–16
mechanism approach to neuroscience, 120–1, 129
memory
 accuracy of, 72–3
 "flashbulb" variety of, 72
 iconic memory and, 49–50
 incorrigibility of, 72
 navigation and, 90, 99
Merleau-Ponty, Maurice, 20, 124, 169n. 63
mesh argument, 3, 41, 48, 50–1, 53
mesolimbic/dopamingeric system (ML-DA), 91, 102–4, 107, 122, 137, *see also* limbic system
metacognition, 7, 41–3, 45–6, 136, 174n. 16

Index 207

method of reciprocal constraint (MRC), 182–3n. 39
Metzinger, Thomas, 11, 148, 151, 176n. 2, 184n. 4
Mill, John Stuart, 105, 178n. 24
minimal cognitivism, 123, 127
minimal conscious state (mcs), 53
modular embodiment, 25, 111, 123–5, 128–9
Moore, G.E., 67–9, 73–4, 159, 173nn. 9, 12, 174n. 19
motivated learning, 121–2

narrative subject model
  anterograde amnesia example and, 33–4
  Aristotle and, 27, 34
  child abuse memory example and, 31
  chronicles and, 27–8
  consciousness and, 28
  critiques of, 25–6, 29–36
  diachronic personalities *versus* episodic personalities and, 36
  embodied subjectivity and, 26, 33–4
  emplotment, 27–8, 31, 35–6
  epistemic/pragmatic relations and, 35
  error through misidentification and, 29–33
  first-person narrative and, 28, 34, 37
  from-the-inside experience and, 29, 31, 35–7
  "I shot J.R." example and, 30–1
  identification free self-reference and, 29, 31–4
  ipse identity and, 28–9, 35
  narrative "I" pronoun use and, 26, 29, 31, 33–5
  narrative self *versus* embodied subject and, 32–4
  P* predicates and, 29–32, 35
  phenomenology of the first-person and, 29, 33–4, 37
  Rovane's Alternative and, 32–4
  situated agency and, 35, 37
  strong narrativism and, 26, 32–5, 37, 168nn. 34, 41, 170n. 64
  weak narrativism and, 26

natural selection, 55, 115–16, 119–20, 136–7, 158, 162
naturalism, 84, 86, 158–62, 164
naturalistic fallacy, 158–9, 186n. 4
navigation
  affective neuroscience perspective on, 103
  cognitive maps and, 4, 90, 95, 99, 107
  fine odor discrimination and, 93–4
  first-person perspective and, 84, 98, 100, 107, 135
  limbic system and, 93, 103–4, 107–8, 113, 136–7
  memory and, 90, 99
  neighborhood avoidance example and, 57, 107, 138
  olfactory system and, 94–7, 137
  subjectivity and, 3–4, 88–90, 93–4, 98, 100–1, 108, 110–11, 113–14, 118–22, 135–6, 138, 177n. 8
  temporality and, 6, 135
neocoritcal plasticity, 89, 184n. 2
neural correlates of consciousness (NCC) research
  binocular rivalry paradigm and, 5, 142–3, 146–52, 156
  content consciousness and, 141–2, 144–5, 147–8
  creature consciousness and, 141, 144–6
  enactivism and, 148–9, 152–3
  explanatory gap problem and, 155–6
  holism of conscious content approach and, 144, 148, 153–4, 156
  inner psychophysics and, 142–3, 145, 152
  intrinsic experientiality and, 145–6
  inus conditions and, 150–1, 153, 156
  manipulationist approach and, 150–4, 156
  matching content assumption and, 146, 148
  mutual manipulability and, 153, 156
  phenomenological critique of, 140–1, 143–9, 153–4, 156

neural correlates of consciousness
 (NCC) research – *Continued*
 sufficiency requirement and, 141,
  144, 146–8, 150–3, 183–4n. 2
 total NCC *versus* core NCC and,
  147–8, 151, 184–5n. 5
 transcranial magnetic stimulation
  (TMS) and, 152
neural plasticity, 89
neurocognitive behaviorism, 102
neurophenomenology, *see also*
  neurophenomenology®
 autopoiesis and, 90, 124
 leib-korber relationship and, 124, 129
 temporality and, 111
 Thompson's articulation of, 7
neurophenomenology®, *see also*
  neurophenomenology
 dynamic systems theory (DST)
  model of temporality and, 133–5
 enactivism and, 134
 functional decomposition and, 134
 goals of, 132
 Husserl and, 132
 Liverpool tides example and, 134
 method of reciprocal constraint
  (MRC) and, 182–3n. 39
 now-phase of consciousness
  and, 134
 reverse engineering and, 134
 temporality and, 134–5
neurophilosophy
 coevolutionary approach in, 157,
  179n. 1
 constraints on, 158–60
 Darwinian naturalism and, 158–62
 genealogical approach in, 158–60,
  164
 genetic fallacy and, 158–60, 186n. 5
 naturalistic fallacy and, 158–9,
  186n. 4
 philosophy of neuroscience
  contrasted with, 157
 of subjectivity, 158, 163
Nietzsche, Friedrich, 159
Noe, Alva, 140–1, 145–6, 148–9,
 167n. 20
nucleus accumbens, 92, 103, 122
Nussbaum, Martha, 59–60

objectivity, 8, 63
old-mammalian brain, 92, *see also*
 limbic system
olfactory bulb, 91, 93–4, 96, 137,
 181n. 16
olfactory spatial hypothesis (OS)
 cognitive maps and, 95
 detector/predictor distinction and,
  94–5
 fine odor discrimination and, 94
 neuroanatomy of olfaction and, 94
 olfactory bulb size and, 94
 parallel map theory and, 95–7
 primary affective modes and,
  103, 108
 prior learning and, 96–7
 space-time loop and, 98
olfactory-limbic system, 91, 93–4
Open Question Argument
 (Moore), 159
optogenetics, 185n. 10
orientation, *see* navigation
outer psychophysics, 142–3
Owen, Richard, 116–17

pain, 3, 13, 30, 155
pair bonding, 17, 117
Panksepp, Jaak, 58, 100–8, 114, 122,
 152, 176n. 1, 178nn. 19–20,
 179n. 25
paradox of subjective duality, 61–70,
 73, 76, 78
parahippocampal gyrus, 91–2
parallel map theory (PMT), 95–8, 100,
 108, 177n. 13, 178n. 18
Pennartz, Cyriel, 138
perception, *see also* awareness;
  consciousness
 empiricism and, 68
 esse is *percipi* and, 67, 69, 78
 gist perception and, 49
 sensations and, 68
perceptual demonstratives, 17, 79
peripheral awareness, 44–5, 171n. 9,
 172n. 23
permanent vegetative state (pvs), 53
Perry, John, 35
Personalism, 162
personhood, 84

phenomenal concepts strategy, 76–9, 176n. 28
phenomenal consciousness (p-consciousness), 41–4, 47
phenomenal overflow, 49
phenomenal unity of conscious content question, 154–5
phenomenology, *see* neurophenomenology®
piriform gyrus, 93
Place, U.T., 174n. 21, 180n. 5
place-preference, 103
Plato, 160
pleasure principle, 107
population thinking, 115–19, 122, 125
positionality, 164–5
P* predicates
  narrative subject model and, 29–32, 35
  subjectivity and, 13, 15, 18–20, 22
predictive coding, 126–7, 135–7
prefrontal cortex, 114
prereflective self-consciousness (PRSC), 131, 162, 182n. 33
"primal impression" (Husserl), 129–30
primary affective modes
  affective neuroscience perspective on, 101–3
  olfactory spatial hypothesis (OS) and, 103
  psychoanlaytical perspective on, 107
  SEEKING as example of, 103–5, 107–8, 137, 179n. 25
  unconscious elements of, 104–7
Prinz, Jesse, 50, 52, 75
problem of unreportable experience among animals, 106
  backmasked stimuli and, 49–50
  dual-task paradigm and, 49, 51
  feature binding and, 49–50, 52
  minimal conscious state and, 53
  phenomenal overflow and, 49
process homology (homology of function), 119, 137
protention-retention (Husserl), 6, 129–31, 133–6

psychoanalytic model of unconscious thought
  depth psychology and, 56, 59
  emotion and, 57
  representation concept in, 55–6
  unconscious subjectivity and, 41, 54–9
psychoanalytic unconscious (Freud), 39, 56, 107

qualia
  coffee cup example and, 63–4
  definition of, 62–3
  direct awareness of, 6, 13, 61, 64, 70
  functional premises and, 112
  immediacy intuition and, 63–4, 69–70, 73, 173n. 14
  incorrigibility of first-person relation to, 70–3, 78
  inner awareness of, 13, 61, 64, 66–73, 76–80, 162, 172n. 7, 175n. 27
  phenomenal concepts strategy and, 76–9
  reality of, 67–8
  sensations and, 68
  as "Someone Else's Problem," 74–5
  subjectivity and, 6–7, 13, 24, 44, 61, 64, 70–1, 76, 80
  tapestry example and, 75
  transparency thesis and, 73–6, 79, 174n. 19, 175–6n. 28

reverse engineering, 115–16, 118, 125, 128, 134, 153
Ricoeur, Paul, 168nn. 37, 41, 169n. 63
Rorty, Richard, 38, 59, 71–2
Rosch, Eleanor, 132
Rovane, Carrol, 32–4
Russell, Bertrand
  on perception and sensations, 68
  Russellian Acquaintance model of subjectivity and, 63, 65–7, 75, 173n. 14
  theory of descriptions of, 14, 18

salience landscape, 88, 165
Sartre, Jean-Paul, 27, 36
Schechtman, Marya, 36, 170n. 64

Schenk, Françoise, 95
Schiff, Nicholas, 53
Schleiermacher, Friedrich, 162
scientific image, 83–6, 158, 164–5, 176n. 2
SEEKING (primary affective mode), 103–5, 107–8, 137, 179n. 25
the self, 165
self-representationalism, 39, 41, 45–7, 76–8, 175n. 27
Sellars, Wilfrid, 78, 83–6, 158, 164–5, 175–6n. 28, 176n. 2
sensorimotor subjectivity (Thompson), 7
septal nuclei, 93
Sheinberg, David, 142–3
Shoemaker, Sydney, 13, 18, 20, 22–3, 31, 147, 185n. 5
situated agency, 35, 37
situated subjectivity, 20–1
Sketch Map, 95–6, 108
Smart, J.J.C., 133
Space of Reasons, 19–20
space-time loop (Freeman), 97–8, 108–9
Sperling, George, 48
Spinoza, Baruch, 86, 104, 178n. 22
Spivey, Michael, 127, 155, 184n. 2
Strawson, Galen, 36, 170n. 66, 173n. 11
striatum/nucleus accumbens, 92, 103, 122
subjective salience, 137
subjectivity, *see also* baseline subjectivity; for-me subjectivity; narrative subject model; unconscious subjectivity
  affective engagement and, 88, 122
  attention and, 51–2
  autopoiesis (self-organization) and, 90
  awareness and, 19, 40–1, 65–6, 131
  awareness of qualia and, 6–7, 13, 24, 44, 61, 64, 70–1, 76, 80
  cognitive maps and, 162
  concept location project in philosophy and, 160–2
  consciousness and, 2–3, 6–7, 39–42, 44, 46–7, 52–3, 55, 60, 72, 88, 114, 145, 165
  coping and, 4, 7, 19, 84, 87–9, 93, 110, 121–2
  Darwinian naturalism and, 84, 86, 158–62, 164
  dreams and, 70–1
  as a "dubious concept," 160–3
  eliminativist approach to, 40, 163
  embodiment and, 1, 4, 110–11, 165
  emotion and, 42, 57–60, 71
  evolutionary developmental biology (evo-devo) and, 8, 93, 110, 118–21
  explanatory problem of, 11
  first-person perspective and, 2, 4, 11–12, 20–1, 29, 47, 52, 61, 67, 70, 88, 91, 108, 110, 114, 146, 166n. 11
  genealogy of, 158, 160, 162, 164
  "I" pronoun and, 12, 23
  identification free self-reference and, 3, 6–7, 12–20, 110, 113, 131, 135, 138, 163–5
  immunity to error through misidentification and, 3, 12–14, 29, 34
  incorrigibility of, 70–3
  intersubjectivity and, 17, 86
  as "intrinsic glow" in consciousness, 40
  Jackson's Constraint and, 111–14, 124, 138, 179nn. 2–4
  joint attention and, 17
  limbic system and, 91, 100, 113
  navigation and, 3–4, 88–90, 93–4, 98, 100–1, 108, 110–11, 113–14, 118–22, 135–6, 138, 177n. 8
  neurobiological image of, 6, 83–6, 113, 121, 124, 129, 164
  neurophenomenology and, 131
  neurophilosophy of, 158, 163
  P* predicates and, 13, 15, 18–20, 22
  pain and, 3, 13, 30
  pair bonding and, 17
  "paradox" of, 61–70, 73, 76, 78
  personhood and, 84
  phenomenal concepts strategy and, 76–9
  phenomenology and, 5, 88
  positionality and, 164–5

subjective salience – *Continued*
  prereflective self-consciousness (PRSC) and, 131, 162, 182n. 33
  reflexivity and, 7
  robot vacuum cleaner example and, 4, 110, 119
  salience landscape and, 88, 165
  self-representationalism and, 39, 41, 45–7, 77–8
  as a "sense of where you are," 164–5
  situated subjectivity and, 20–1
  spatiotemporal articulation of, 24, 88, 90, 136
  temporality and, 5–6, 110–11, 119, 129–36, 138–9, 167n. 22, 182n. 33
superblindsight, 45–6
superior temporal sulcus (STS), 143, 149, 156
supervenience, 105–6, 152–3, 178n. 24
Syal, S., 121–3

Teichert, Dieter, 26, 28–9, 168n. 37, 170n. 64
temporality
  Husserl on, 5–6, 111, 182n. 33
  navigation and, 6, 135
  neurophenomenology and, 111
  protention-retention structure (Husserl) and, 6, 129–31, 133–6
  subjectivity and, 5–6, 110–11, 119, 129–36, 138–9
thalamic nuclei, 92
thalamus, 92, *see also* hypothalamus
Thompson, Evan, 6–8, 90, 123, 128, 131–2, 140–1, 145–6, 148–9, 153–4, 167n. 20, 182n. 36, 186n. 7
Tononi, Giulio, 91, 184n. 5
topological maps, 100
transcendental phenomenology, 124, 162
transcranial magnetic stimulation (TMS), 152
transparency thesis, 73–6, 79, 174n. 19, 175–6n. 28
triune brain hypothesis, 92
Tye, Michael, 21, 74–6, 143–4, 154–5, 175–6n. 28

unconscious subjectivity
  affective neuroscience and, 57–8, 104–8
  cerebral unconscious and, 57
  cognitive model of unconscious thought and, 41, 54–5, 57–8, 60
  consciousness contrasted with, 38, 41, 44, 47, 55
  emotion and, 42, 56–60, 71
  first-person unconscious and, 56–7
  for-me mental states and, 48, 54
  Freud and, 39, 54, 56
  iconic memory and, 50
  Kriegel's argument against, 7
  in me/for-me distinction and, 39, 54
  mesh argument and, 3, 41, 48, 50–1, 53
  minimal conscious state and, 53
  neighborhood avoidance example and, 57, 107, 138
  phenomenal consciousness *versus* access consciousness and, 41–4
  problem of unreportable experience and, 47
  psychoanalytic model of unconscious thought and, 41, 54–9
  self-representationalism on, 41
  zombie subjects and, 171n. 11

Varela, Francisco, 128, 132, 148–9, 153–4, 181n. 26, 182n. 31
ventral striatum (VS), 91, 103, 108, 122, 128, 136–8
ventral tegmental area (VTA), 92, 103
visual consciousness, 52, 145, 149, 184n. 3
vocal learning, 121–2

Wimsatt, William C., 120
Wittgenstein, Ludwig, 12–13, 18
Worldly Experience, 19

Yeats, William Butler, 79

Zahavi, Dan, 129–32, 181n. 28, 182n. 33

GPSR Compliance
The European Union's (EU) General Product Safety Regulation (GPSR) is a set of rules that requires consumer products to be safe and our obligations to ensure this.

If you have any concerns about our products, you can contact us on

ProductSafety@springernature.com

In case Publisher is established outside the EU, the EU authorized representative is:

Springer Nature Customer Service Center GmbH
Europaplatz 3
69115 Heidelberg, Germany

www.ingramcontent.com/pod-product-compliance
Lightning Source LLC
Chambersburg PA
CBHW071614100426
42873CB00004B/46